Divine Evolution

a hybrid theory to reconcile creationism and evolution

John L Leonard

Published by Each Voice Publishing

www.eachvoicepub.com

ISBN: 061558652X
ISBN-13: 978-0615586526

Library of Congress Control Number: 2012930258

Editor(s): Donna Sundblad, Sharon L. Connors, Lisa H. Leonard
Front Cover Design by Lisa H. Leonard based on photography by Kathryn Williams

First edition, 2010, Second edition, 2012

PERMISSIONS

Lyrics in Chapter 1:
MONKEY SCHEME
Lyrics by J. John Jackson

Lyrics in Chapter 9:
From the Rolling Stones for Shine a Light
Words and Music by MICK JAGGER and KEITH RICHARDS

Lyrics in Chapter 10:
Highway 61 Revisited
Written by Bob Dylan

DEDICATION

This book is dedicated to the glory of God and the quest for knowledge.
Professional dedications go to:
My loving, eccentric, charismatic, and infinitely entertaining extended family and my friend Donna for recommending the book to her publisher.

To my mother, who encouraged my voracious appetite for reading.

Last, but never the least, to my wife Lisa, editor second to none. Her faith in me and encouragement allowed me the freedom and flexibility to sit down one day and let years of inspiration, ideas, and contemplate be committed to prose.

CONTENTS

ACKNOWLEDGMENTS

To Mike, for sharing my book with his mother, which got me connected with my writer's guild and started the ball rolling.

To WAG (Writer's Alliance of Georgia) for helping me improve as a writer via constructive criticism.

1 GENESIS

(The following is sung to the Monkee's theme song)
They say we were monkeys
I think that they've been monkeyin' around
Cause they're too busy changin' their books and their studies 'round
They're just tryin' to defend it—they got a lot to re-explain
Where they've done excavation, they keep uncoverin' mistakes

Any time or anywhere—just look over the boulders
You'll see that man's been there
They say we were monkeys
I think their heads are stuck in the ground
Cause they're too busy diggin' and putting the Bible down

- J. Jackson [1]

Science has accomplished many wonderful things to the great benefit of mankind. Now if we could only do something about the plethora of lazy scientists today, we would surely be living in a better world.

More than once I've been asked why I felt the need to write this book. It's a good question, one I've asked myself a time or two. The truth is after I started the process, I couldn't stop. The many years of cumulative frustration from listening to pseudo-scientist talking heads pontificate their wildest theories without any serious challenge finally bubbled to the surface. I found myself asking, "Isn't somebody going to speak up and ask a serious question or two? Does anyone else here even notice the Emperor has no clothes?"

Steven Colbert is to blame as much as anyone. It was his program on television playing in the background while I was working on a website for a new business. Colbert was interviewing Richard Dawkins, author of *The God Delusion.* Dawkins made the most ridiculous comment I had ever heard on television. I stopped working, found the remote control and backed up the live satellite feed (TIVO is wonderful) and listened to it again. I backed it up a third time and recorded the playback so I could listen over and over. No matter how many times I listened, I couldn't believe what I was hearing.

Each time I heard Richard Dawkins say that cars, computers and cell phones were intelligently designed but humans were not, my head wanted to explode. Was this guy serious? At the same time, a more relevant question was, who is he?

After a quick Google search, I found out Dawkins is a world famous zoologist and author of a number of books. It turns out he may be the most famous atheist in the world today. I admit I disagree strongly with Dawkins, so you'll probably figure I'm an extremely devout Christian who goes to church three times a week and prays every hour. You would be wrong.

As I said at the beginning, this is Colbert's fault. If he'd asked a few tougher questions of Richard Dawkins and challenged his assertion against intelligent design, I probably would have gone back to working on the website. And where was Colbert's show? On television. If you read the lyrics to Frank Zappa's song *I'm the Slime,* you'll realize they are very relevant considering how this all got started.

Besides, who ever accused me of being a good Christian? I'll be the first person to admit I call myself a Christian and I believe what I believe, but I'd never suggest I'm very good at it. I go to church, but certainly not every Sunday. When I lose my temper I could put the saltiest sailor to shame. I can even utter curse words in foreign languages.

Oh yeah, and I like Frank Zappa. Nobody's perfect.

Four chapters of this book begin with song lyrics quoted from a favorite tune. Originally, in my first draft, every chapter of the book started with a song lyric. At the time, I had no idea I was creating a nightmare for myself in terms of copyright use permission violations.

I wanted to open the book with a verse or two from Zappa's song, but his heirs at Munchkin Music declined permission to reprint the lyrics. The song fit nicely, and mostly served to demonstrate my open-minded view of music, literature and information in general. I do consider the source, but don't automatically exclude information because of it. Just because Frank Zappa could be gross and disgusting at times did not mean he never said anything intelligent or insightful. And no matter what words came out of his mouth, the music from his guitar was nothing short of awe inspiring.

I was already using J. Jackson's lyrics in the next chapter and had reprint permission, so I moved them to open the book. My misinterpretation of the Fair Use clause regarding copyright use permission nearly set up a potential litigation nightmare; I thought if only a line or two of a song was used, it qualified as fair use. Songs are treated differently, I found out. The final solution turned out to be rather than take any chances, if I'd already asked for permission and received it, the lyrics were retained. Otherwise they were deleted.

Before the book became a commercial endeavor, it was a labor of love (for knowledge) that I printed at home and distributed to friends in a loose-leaf notebook. The first thing that happened after a publisher offered a contract was it created a mad scramble to gain copyright use permission. The lyricists who wrote the songs I quoted from to begin each chapter had to be contacted. I will be eternally grateful (and that's a long time!) to Bob Dylan, Mick Jagger and Keith Richard, and J. Jackson for the permissions granted to quote the lyrics used within.

I can't help the fact I think Frank Zappa was funny and played great lead guitar. I also think *Monty Python, The Black Adder, Green Acres*, and *The Beverly Hillbillies* were funny so maybe there's just no accounting for my sense of humor. Perhaps I may find a need at some point to stop writing long enough to look for some zircon encrusted tweezers…but I digress.

Consider the reference to Frank Zappa as a gentle advisory warning. I'm certainly not trying to drive away readers, but I have to be honest as the book's original title proclaimed I would. Because that original title was kind of stupid, we'll just leave it with the acknowledgement that the word "honest" was in that working title.

It's entirely possible I will alienate the atheist and the evangelical Christian in one fell swoop. Neither is a goal. To the atheist, I'd like to offer a perspective you may not have found elsewhere. My perspective is one of a person who started as a child with childish faith. I have run the gamut to the very brink of atheism and returned.

I later dismissed that faith as fiction as my education progressed. Simply put, I was taught to know better. Personal experience and my journey through life led me full circle to believing in God again, though not with a childish and immature faith but a faith now tested by fire.

So I ask the atheist reader to bring an open mind to this discussion. Once upon a time we shared a similar point of view. I will present a broad collection of information and ideas for your consideration, and I only ask that you fairly balance and weigh the evidence and decide its validity after you know what it is, not beforehand.

I also ask the Christian reader to keep an open mind, as I will provide what I believe is compelling evidence to consider an alternate perspective regarding some of the entrenched beliefs in "mainstream" Christian religion without letting go of their faith. My views may differ from the views of other Christians, particularly in the areas of how I interpret the meaning of the six days of creation and my estimate of the Earth's age. Please listen to what I have to say and judge it for what it's worth, which is no more or less than any other scientific theory regarding these questions surrounding the truth about the origin of life.

TYPICAL JOURNEY

I was once taught the Earth is young. I believed it until scientific evidence challenged that belief with more plausible data about its age. For years, I struggled with my inability to reconcile the religious teachings that provided answers for these questions with what science taught me. It led me to reevaluate my faith in God and through a long period of serious doubt, if not disbelief in Him.

I eventually worked things out in terms of finding a base point of my belief system (I had a lot of help). I read a book by an astronomer named Dr. Hugh Ross called *The Creator and the Cosmos,* which gave me a good bit of scientific information about the Big Bang Theory but from the context that God was responsible for its occurrence. It was as if someone turned on a bright light in a dark room. But that was years after my personal experience had already led me to believe in God again and created the constant base point of my faith.

After I graduated college and began to live my adult life, until I had my personal epiphany about God, I realized the gaps between what we think we know and what is true regarding the theory of evolution are actually gaping holes. Science claimed these gaps would be filled eventually by knowledge and understanding, but these holes were big enough to drive a truck through. I had questions, and I started looking for answers. Surely questioning the known is a positive growth experience about which Dawkins and I may agree.

The Christian reader has been told for years by dismissive evolutionists that he or she believes in a mythical God no more real than Santa Claus or the Easter Bunny. Science declared victory over religion based on a couple of points in evolutionary theory proven true in the last two centuries. Well, at least they seem to be true.

We can only go where the evidence leads us. The evidence indicates the arguments for evolution which appear to be true also support quite nicely arguments for believing in an Intelligent Designer. Conversely, there are points in evolutionary theory that fall apart completely without God's presence. We are told by our evolutionist friends that those parts of the theory of evolution simply "need some more work." That is the understatement of a lifetime.

God doesn't really need me to defend Him. I am defending my belief and faith in my Creator. I am attacking parts of a scientific theory elevated to factual status when the evidence clearly does not support doing so. In fact, my first step in the next chapter will be to establish that there is no such thing as a "Theory of Evolution." Four different scientific theories are often called "evolution" or used interchangeably with the term, but none of the four theories may truly be accurately called evolution.

While I intend to provoke thought in my evolutionist friends, I do not wish to provoke anger. After all, I know how it feels. I've been called delusional (and worse) by Richard Dawkins because of what I think. It is difficult to endure obnoxious insults from the mouth (or typewriter) of someone whom I don't even know. So I will try my best to refrain from responding in kind with the use of inflammatory rhetoric to describe people with contrary opinions. There is a difference between calling a person stupid and pointing out the person said something stupid. We all say stupid things occasionally, and when Richard Dawkins happens to take his turn, I will endeavor to minimize any glee I display when I point it out.

The first to lose his temper and shout insults is the person who has lost the debate. It's the favorite ploy of modern society—when in doubt, shout them out. Civil discourse is rapidly becoming a thing of the past.

After reading *The God Delusion,* my first thought was "Are those the best arguments you have?"

Then I realized what was missing. The book wasn't really a defense of evolution; it was an assault on religion. It seemed that Dawkins' book was awfully short on facts and data supporting evolutionary theory but incredibly long on the wild rants against organized religion. Given his professional training, I was incredulous that Dawkins refused to admit what he surely knew about the human brain. The man is a zoologist but chose to ignore those inconvenient truths (now there's an ironic choice of phrase).

CONSIDER YOUR BRAIN

The children's book *Ripley's Believe It Or Not!* shows the human brain may only weigh on average a little over 3 pounds, but it is "the most complex collection of matter in the universe. The total memory capacity of the brain is estimated at 100 billion bits of data, which is equal to 500,000 large multi volume encyclopedias, or in computer terms, about 1,000 gigabytes." [2] That's approximately one terabyte of data. And we don't even use more than half of the brain God gave us.

The human brain surpasses the capabilities of the greatest computers man has ever built. Computers are nothing more than tools that handle instructions provided by human programmers. There's nothing all that smart about them. Turn off the power or remove its operating system, and a computer is less useful than a hammer or screwdriver. It might be useful as a cumbersome paperweight or perhaps as a doorstop.

While the processing speeds achieved by computers surpass the mathematical capabilities of the human brain in terms of sheer speed, at the end of the day they are only at best really fast calculators. Computers are limited to being as good as the accuracy of the data they process and the competence of the programmers who wrote their instructions. The old programmer's term to describe the limitation is GIGO, an abbreviation for Garbage In, Garbage Out.

By comparison to artificial intelligence, the abilities of the human brain are virtually limitless. Consider the work required to accomplish the following task: walking across a room. The room in which I happen to be sitting has the normal obstacles one might expect in an office. Desks, chairs and a few other pieces of furniture present fixed or stationary objects. In this particular room, a couple of dogs also keep me company as I write, which create variable or mobile objects as well. If I get out my chair, the dogs will probably stir. So I'll get up carefully in hopes they don't move. I'll cross the room, watching my step as I go. I don't really think about the effort required to walk across the room because the prominent thought in my mind is my reason for crossing the room–I'm thirsty. I'm headed to the kitchen to get a glass of iced tea.

I'm not saying a robot can't do the same thing. What I am saying is that most of what I described are actions we humans don't consider work because we don't have to consciously think about them to do them. The only real command in the conscious mind is "I'm thirsty." Every other activity necessary to accomplish the goal of the work was performed by the subconscious mind.

When I stood and looked across the room, my eyes registered the location of every obstacle, both fixed and variable. With each step I took, I received feedback from the optical processing center in the macula of my eyes. A sub processor controlling my appendage movement directed alterations in my steps to avoid sudden obstacle variations. In other words, the dogs moved. In computers, these variations invariably manifest themselves as "bugs" because of their inherently volatile nature.

Bugs are the failure of the programmer to equip the computer to deal with every possible outcome or variation for a given task. In fact, the original computer "bug" was literally an insect. Somehow, it got inside the box and adversely affected the operation of one of the first computers ever made. The designer of the machine neglected to protect against the possibility of external intrusion into the system by an insect, and catastrophic failure resulted.

While I was rattling on about computer bugs, I successfully navigated the room, and *voila!*–I have my refreshment.

The computerized robot can accomplish the exact same set of tasks, but only if instructions exist to handle both the normal processes (standing, taking steps) and the exceptions (the phone rings, a dog moves, loss of electricity). The computer must process each instruction sequentially in linear fashion. If you happen to leave out an instruction, you instruct the robot to take a step with its left foot before you told it to stand, the robot's computer brain cannot make the leap in logic and understand that you meant for it to stand up first. It requires every direction, step-by-step, in painstaking detail. Implied or obvious and unnecessary commands to a human cannot be assumed for our robot friend. But the robot's instructions can be modified to get the job done.

At least, you can get the robot to cross the room and bring back a soda if that's what you want it to do, as long as you write the code and handle every exception caused by both foreseen and unforeseen variables. As long as the program handles every exception gracefully, the computer won't crash.

But that only gets you a drink. Now let's consider the other processes running in the background that humans get for free. The phone rings in another room. For the human, an auditory listening process running as a background agent signals the brain literally that there's an incoming call. An interrupt is signaled to the ambulatory processors with a new message: change direction. Now go answer the phone.

Mission accomplished for the human. But I suspect we just broke our robot.

For a human being to do any of this is not a major accomplishment. We don't consider it work to go get something to drink for ourselves. The human brain is a central processing unit (CPU) far superior to any made of silicon, but what really sets our abilities apart from machines is our distributed processing capabilities. The sub-processes of the subconscious allow the human brain to dust its closest competitor because we get all this processing capability just by being born human.

A big deal was made when Big Blue, the IBM supercomputer, beat world champion Garry Kasparov in chess. But the first time they played Kasparov won. The computer was taken back to the drawing board by its team of programmers and was able to win the second time. Yet the computer wasn't able to think about anything but chess during the game; it could only execute its programmed functions. Kasparov could think about world politics, or what to eat for dinner; he might think about them, or he might not. He has the choice, or maybe he doesn't.

Haven't you ever gotten a song "stuck" in your head? The mind is a powerful thing and sometimes is difficult to control, even by its "owner". Big Blue was incapable of thinking of anything not programmed into its software instructions, which ostensibly would only be information about chess given its task to beat the world's greatest human chess player. Big Blue is a chess savant.

OPEN MINDS

I don't want to be presumptuous and start sounding like another Richard Dawkins. I do not arrogantly pretend to have all the answers. I'm not even sure I've thought of all the right questions, but I did come up with some good ones and you might want to ask them yourself as well.

I hope you will be open-minded about the possibilities I offer for your thoughtful consideration. This book will not make any absolute statements about faith because I don't believe I am absolutely right. I only believe what I believe and for the reasons I'm about to give. My faith continues to evolve, and that is evolution I can believe in.

Before I wrote anything critical of evolutionary theory, I felt it best to make sure I was familiar with the latest and greatest information on the subject before I said or wrote anything, particularly after I realized that Dawkins had so little to say about evolution in his own book. I will always remember an old quote attributed to Abraham Lincoln, "It is better to remain silent and thought a fool, than to speak up and remove all doubt."

Ronald Reagan was once asked about signing a nuclear arms treaty with the Russians as to whether or not he should trust our Cold War enemy. Reagan answered "Trust–but verify." Those were astute words then and remain so today. I learned the truth of them the hard way from my dad in the form of a practical joke he played on me once.

During dinner, he handed me a hot dog prepared with mustard and ketchup smothering the frank just the way I liked, and I viciously bit into a rubber hot dog. Because my dad prepared the hot dog and gave it to me, I trusted him blindly. I bit hard, and didn't just give up immediately when my teeth failed to puncture the rubber hot dog. It took several seconds to realize the rest of my family was laughing at me because I didn't stop trying to bite through the rubber. It never fully registered that I was the victim of his prank until he took the fake hot dog away and gave me a real one.

I was probably only about eight or nine-years-old at the time, but I have never forgotten the lesson I learned: look before you leap. Trust but verify. I think that incident helped build the foundation of my skeptical nature. From that point forward, I've always had problems accepting information at face value.

In part, that is why I took so long to research this book. I was not satisfied merely to say "well, they're wrong" and be done with it. I'm not wired that way, and that's a bit of a lazy way out. If you want to be taken seriously, you can't just say "he's wrong" or "I'm right" without substantiating your reasoning.

I still wonder about the existence of life on this planet. Even though I believe God did it, I'd still like to know *how* He did it. Though I cannot conceive of drifting away from my faith again, I cannot help but remain interested in contemplating answers for the existential questions.

Consider Stone Mountain, in the outskirts of Atlanta, Georgia, which is actually too small to be categorized a mountain. Technically Stone Mountain is a big hill. Yet, it is dubbed the largest mass of exposed granite in the world, or at least that is a disputed claim made by local tour guides who work at Stone Mountain Park.

The tallest manmade structures in the world on September 11, 2001 were the twin towers, but jet airplanes flown by terrorists caused both to fall. Had both planes flown into Stone Mountain instead, the damage to the "mountain" would have been mostly cosmetic. Some human development might have been destroyed, but Stone Mountain would remain standing. The loss of life on the planes would still be tragic, but the catastrophic loss from the towers would not have occurred.

Nuclear weapons could probably destroy it, but conventional weapons would barely take a chip off Stone Mountain. That monolithic block of granite is so heavy and massive it boggles the mind, until you consider the weight of a real mountain like Mt. Everest or Mt. Fuji.

Think about it. Then consider the fact that Stone Mountain, Mt. Fuji, the entire chains of the Andes, Rockies, and Himalayan mountains are nothing–a couple of pebbles when compared to the size and scope of the entire universe. And every bit of matter on our planet and throughout the universe had to be involved in the Big Bang. It's a matter of scale and perspective.

We tend to be more impressed by "chunneling" under the English Channel or the world's longest suspension bridge and miss the whole point of the real wonders in the world there under our noses. Is the Golden Gate Bridge in San Francisco really more impressive than the landscape it connects or the bay it spans?

To examine the complexity and obvious design of that remarkable creature we call a bat opens the door to arguments about irreducible complexity that we will address in further detail later. The fact that human inventions we believe so marvelous are designs stolen from natural models is lost on the majority of people. The sonar, which has proved so invaluable to the Navy, assists a ship to navigate through treacherous waters safely. The device was designed after some inventor observed of how bats and dolphins use sound to echo-locate objects in nature, leading to "his" clever idea.

Our solar system is dwarfed by our galaxy, which is one of many in our massive universe. Imagine there's a giant map of all we know of space. Picture a small mark on the map representing earth. The mark has a microscopic little red dot on that earth mark with a note that says "you are here." This may help your perspective. [3]

As we begin our debate, we will identify the four components of evolutionary theory and the evidence for each. We will discuss the fossil record, the available evidence that supports Darwinism, and the current status of each of the four components of evolutionist theory.

My faith comes from a skeptical foundation, and I will not hold back any information, public or personal, on how I came to have that faith. My faith is a strong enough that it makes me want to publicly challenge a prominent author and well respected academic on the core message of a book he wrote.

This book directly attacks Dawkins' assertion that God is a figment of our imagination. I will provide ample logic and apply common sense to validate my reasoning. We will share the same evidence. We will argue over the meaning of the fossil record. I will use Dawkins' own words and respond to them point by point.

I have separated evolution into the four real theories that are each mistakenly lumped together as **Evolution** with a big "E" so we can deal with each individually. I will concede the points with which I agree and contest those meriting further discussion. There are some instances where science and religion could both actually be right. You may be surprised when you ponder the full body of evidence I will present.

But let me be clear…where scientists denigrate religion and insist God is not possible is exactly where their arguments fall apart. I will make every attempt to identify, discuss, and refute those arguments and the flawed reasoning that spawned them. But in a kind and gentle fashion, of course. And remember–if you don't like this book, blame Stephen Colbert.

2 WHAT IS EVOLUTION?

I believe in evolution. It's the questionable set of scientific theories posing as **Evolution** that bother me so much.

CHANGE IS INEVITABLE

Evolution simply means change. Change occurs around us every day; it's easy to believe because we observe it. In fact, we ourselves evolve. At least our consciousness does. My view of life has certainly evolved as I have lived it. To paraphrase Winston Churchill, when I was twenty I had a heart. When I turned forty, I also had a brain. In between those age milestones, I gradually came to wonder if God was any more real than the Easter Bunny. I followed the very path toward atheism Richard Dawkins hoped I would for a while.

Of course, technically speaking, I always had a brain. I simply had not used it very often to think about life, the universe and everything. [4] Those things didn't seem as important as career advancement, or what time my next tennis match or golf tee time was scheduled.

Things changed, and I have changed. I prefer to think for the better. In his book, Dawkins says his goal is to show religious people with open minds they can find happiness outside God. By contrast, my goal for this debate is to demonstrate to both secular and religious people there is evidence in science and beyond pure science to justify belief in God. Also, the best and most truly convincing evidence supporting evolutionist theory coordinates quite nicely into a slightly modified version of creationism belief that I call Divine Evolution.

Heck, even the theory of evolution has changed. Thirty years ago, man was supposed to have descended from Old World Apes. Now with DNA evidence apparently contradicting the Old World Ape theory, humans are said to be descended from an undetermined third form of chimp according to the latest most popular theory. And if the real truth about the comprehensive theory called Evolution were known, you will have to eventually believe that our most distant ancestor is not a chimp, but probably either an oak tree or perhaps just some pond algae.

Thirty years have elapsed since my last classroom instruction regarding the theory of evolution. I knew about Lucy (*Australopithecus*) and the fossil record. I studied the evolution chart, progressing from ape to man years ago. I learned about the Big Bang Theory. However when I looked at the big picture, I saw lots of problems with the purely scientific explanations.

I drew my own conclusions. The theory of evolution contained interesting concepts and little hard evidence to back it up. I also had problems accepting the young Earth theory from strict creationist teaching. Some evidence supported a belief in an old Earth in the form of old bones, but when I looked at a chart picturing an ape slowly transforming into a man; quite frankly, it made me laugh. The idea seems so ludicrous, not to mention we have both apes, chimps, and man. If man had truly descended from apes or chimps, why wasn't the ancestor species completely out competed?

When I looked at recent revisions of the same chart, I laughed again. Cro-Magnon Man looked remarkably similar to a Robinson Crusoe-like Tom Hanks in the movie *Cast Away.*

As I understood Darwin's theory of natural selection, applied within the limits of the species boundary, it made perfect sense. When extended beyond the species boundary to provide a mechanism to make feasible the theory of speciation, natural selection made no sense at all.

Natural selection explains how we get brown bears, black bears, and grizzlies; speciation is the theory that attempts to explain how we get alligators, deer, and vultures. From the theory of the origin of life to the development of every form of modern life, speciation is the alleged mechanism that accomplishes that diversity outside of God shaping the individual species. This means humans and crocodiles are distant cousins, but when I look at one, I fail to see the resemblance.

Of course, there was hard evidence, which also brought into serious question a few things I'd been taught about the Bible as a child. Somebody in a church somewhere added up the chronology given in the Bible of the genealogy from Adam to Jesus, tossed in six extra earth days for God to create everything and *voila*, the age of the Earth was projected to be between 6,000 and 10,000 years old. Clearly, there is abundant evidence in the form of the fossil record which contradicts those age calculations. I don't think the dinosaurs lived and died within a few thousand years, less than ten thousand years ago.

My knowledge of science provided enough evidence that made me think the literalist belief in a young Earth was wrong. The age of the Earth is not literally written anywhere in the Bible. It is calculated and taught by extrapolation.

The young Earth theory is not part of any mainstream Christian creed. I also came to believe "my" version of the Big Bang Theory was true. This is not to say I no longer had any questions about an old Earth or the Big Bang, as we will see shortly

SCIENCE AND FICTION

I still believe the young Earth theory is wrong. But I also know much of the evolutionary theory is pseudo-science or pop science, whatever you want to call it. Science is an essential study of the world around us, which leads to medical breakthroughs and seemingly limitless advances in technology.

Pseudo-science is consensus opinion posing as solid work product developed due to the ability of inferior scientists to stand on the shoulders of giants and create their own original work.

We'll keep working to find the right words to convey the meaning we hope to achieve. My favorite candidate for the term, at the moment, is "fiction science" as opposed to science fiction, which we know we aren't supposed to believe is real. Fiction science is pseudo-science passed off as true science when it is not.

My first objective in preparing this debate about evolution was to learn the current meaning of the word, which we have established means "change" according to the dictionary. Darwin's theory of evolution is actually an oxymoronic phrase because his theory is not called evolution. Darwin's theory of natural selection describes intra-species mutation. The scientific theory of evolution, called the Big Bang Theory, explains the origin of the universe from a scientific perspective. Abiogenesis is the evolutionist theory intended to explain the origin of life, and speciation is the theory that attempts to explain the development of multiple species. Finally, natural selection is Darwin's theory, which describes how breeds develop within a species. All four theories are evolution, but none are. It's a neat trick.

Thirty years ago evolution meant apes changed into people by some transformative process. This theory was taught as essentially the way things had to be. Creationism was and is the only alternative theory to evolution, but pseudo-scientists aren't willing to risk creation science gaining equal footing with evolution for a serious debate. Apparently, this is due to fear of losing the debate.

Outside of a course on religion, educational materials routinely exclude the only competitive theory to evolution. This close-minded approach to education is justified by declaring creationist theory isn't scientifically based or it's a constitutional conflict between separation of church and state that violates our freedom of religion as guaranteed by the Bill of Rights.

The creationist is not permitted the same freedom of speech enjoyed by the evolutionist, yet both are governed by the same Bill of Rights. This blatant discrimination is enforced judicially under the pretext of the separation of church and state and freedom of religion are both rights protected by the Constitution.

Freedom of speech is one right apparently deemed inferior to another (freedom of religion) in order to protect evolution science from criticism by using the Constitution. Or maybe the logic is two rights trump one; separation of church and state plus freedom of religion win over freedom of speech. The same freedom of speech sought to teach evolution at stake in the Scopes monkey trial was also at stake in a more recent court case involving informative stickers the county school board wanted put on the cover of science textbooks in Cobb County, Georgia.

Over two thousand parents had complained that the science books used in the local school system taught evolution as fact and not theory. To honor the parent's wishes, the school board voted to put stickers on science textbooks that read "This textbook contains information about evolution. Evolution is a theory, not a fact, regarding the origin of living things. This material should be approached with an open mind, studied carefully and critically considered." [5]

The information on the sticker is entirely correct and accurate. The sticker makes absolutely no reference to creationism. None. There is no logical reason the stickers should have been banned.

But that is exactly what happened. The parents of six children were joined by the ACLU and sued the school board. The same scientists who revere the courts for the Scopes decision that opened the door for evolution into the classroom fought tooth and nail to prevent any alternate theory to be mentioned in class.

On the contrary, evolutionists provided affidavits to support the lawsuit and oppose creation science even one scintilla of consideration. After the Scopes trial, the argument had gone from being that you couldn't teach evolution to you couldn't teach anything but evolution practically overnight. It is incredibly ironic that educators so passionate about expanding young minds in *all* other areas insist any acknowledgement of creationism is dangerous, destructive and downright illegal.

If the federal judge in the Cobb County book sticker case knew the law even as well as I do he would have ruled in favor of allowing the stickers. Creation science would be allowed to compete in the classroom as the alternate theory to evolution science for two reasons. First, the mythical separation of church and state invoked so frequently by anti-religious zealots does not exist as far as any official government document is concerned. The phrase originated in a letter Thomas Jefferson wrote to the Danbury Baptists to assure the Baptists there would not be a national church. Many of the early settlers fled England to avoid persecution by the state-controlled Anglican church and treasured their freedom to worship (or not worship) the God of their choosing above all other freedoms. Separation of church and state is not a role delegated to government by the Constitution.

Second, the Bill of Rights guarantees freedom ***of*** religion, not freedom ***from*** religion. This only means there is no state sponsored church. When thoughtful, reasonable and intelligent judges are appointed or elected to the bench, this inequity will someday be overturned.

SEEKING ALTERNATIVE VIEWS

Before writing this book, I read *The God Delusion* to gain insight into the mind of Richard Dawkins. His book contained surprisingly little information or evidence to support evolution, but that was understandable to a point. Just from its title, one could tell the book was not meant to pass for a primer on evolution. As expected, most of the book argues why the reader should not believe in God.

More argument advocating belief in evolution would have made the book stronger. Nearly 400 pages of blather informing the reader they were stupid to believe in God was a bit much. Apparently, Dawkins assumed his readers knew enough about evolution to already believe it. He focused his energy on offense rather than defense. And he succeeded in being offensive.

Having failed to find the current definition of evolution in Dawkins' book, I became the county library's most frequent visitor during the next year. Still searching for the consensus definition of evolution, I updated my knowledge of the theory by reading *Evolution and the Myth of Creation* by zoology professor Dr. Tim Berra of Ohio State University. Being a relatively short book, I assumed it would quickly get to the point with the best contemporary evidence for evolution. I expected to read about recent scientific breakthroughs and developments in the field.

I was mildly disappointed. After little more than 100 pages detailing the peculiarities of alleles, hermaphrodites, and even more riveting topics like allopolypoidy speciation in plants, Berra suddenly shifted topics drastically. He began to describe the evolution of design as seen in the Chevrolet Corvette.

Throughout Berra's book, I didn't see breakthrough evidence supporting evolution theory Dawkins assumed we all knew was there. Surely evidence existed, which made belief in God such a delusional proposition for Dawkins to so adamantly declare it to be true.

After adjusting to Berra's abrupt change in topic from scientific discussions of heredity and genetics' role in evolution to more fanciful mental exercises in automotive design, I immediately recognized the logical flaws in his analogies. I don't begrudge the man his fun derived from playing with an analogy between a car and a living organism to demonstrate the application of evolution in action. But flaws are flaws, and tough to ignore once they're noticed.

The problem with his examples is they all smack of the obvious need for intelligent design. It is true that Corvette design is "descent with modification" as Berra points out, but the modifications are purposeful, which requires intelligence.

He writes "If you compare a 1953 and 1954 Corvette, side by side, then a 1954 and 1955 model, and so on, the descent with modification is overwhelmingly obvious. This is what paleontologists do with fossils, *and the evidence is so solid and comprehensive that it cannot be denied by reasonable people.*" [6] [emphasis original]

Dr. Berra is absolutely correct to assert evidence of descent with modification obviously exists. Unfortunately for his analogy, what is even more obvious than "descent with modification" is the work of the intelligent designer responsible for the modification. Engineers designed the new car models for the following year with improvements, not regressions in functionality from previous models.

One of the primary arguments that evolutionists make against religious belief is that creationists all believe God created every animal on earth in the exact form of the organism as it exists today. Perhaps some creationists do, but not all.

I have a different theory about God that is essentially a hybrid of what I've learned from science and what I know about God and the world around me. Science has actually helped me better understand my relationship to God in the framework of creation.

Dr. Berra attempts to convince the skeptical reader of the flaws in creation science, only to inadvertently expose his own theory's flaws. Berra asks a series of rhetorical questions and helpfully provides the answers for the reader. One of the questions addresses the problem of entropy, meaning chaos or disorder. Berra writes "*Evolution violates the second law of thermodynamics. Entropy (disorder) is always increasing. Since order does not arise out of chaos, evolution is therefore false*" [7] in order to refute the statement.

To counter the creationist argument for the law of entropy, Berra suggests you can get order from disorder if you apply energy. He uses the example of the receipt of a bicycle separated into multiple parts and packed in a shipping carton. After application of your muscles (energy) the bicycle is assembled. However, more than muscle is required to produce an intelligent assembly of disparate parts. Wildly flailing at the bicycle parts also expends energy, but it doesn't assemble the bike. The application of intelligently directed energy is required to perform work.

Productive work requires intelligent application of energy. The application of your muscles is directed and controlled, while you follow the set of instructions for assembling the bike. If you take the same carton full of parts and dump them at the base of the Niagara Falls, the energy created from the falling water is far greater than the energy expended by your muscles, but which is going to put the bike together faster? No matter how many times you repeat the experiment, the water's never going to assemble a working bike. Entropy or decay guarantees failure or death, not evolution.

The easiest way for an evolutionist to declare victory in a debate about evolution is to wait for the mention of The Big Bang, speciation, or abiogenesis to be described as evolution, then pouncing on the opportunity, evading the question of what evolution is and replacing it with a lengthy discussion of what it is not. Evolution has been used synonymously with natural selection for so long the evolutionist has to make allowances for that mistake from the skeptical questioner, but mention any of the other three and call it evolution by mistake then consider further meaningful debate hopeless.

Ironically, in the Steven Colbert interview, Dawkins utilized that very tactic to defend all of evolutionist theory under the guise of natural selection. When Colbert theorized that if God did not exist, it meant the world was created by a series of random events, at which point Dawkins pounced and exclaimed that Darwinian natural selection was the opposite of random chance. The problem is, Colbert wasn't talking about natural selection, he was talking about evolutionist theory as a whole.

Dawkins turned the conversation to natural selection in order to be able to proclaim "Darwinian natural selection is the exact opposite of random chance. It's a highly non-random process." [8] Dawkins is technically correct with his facts about natural selection but horribly misleading to imply Darwin's theory is an encompassing solution for the origin of the planet or the origin of life.

EVOLUTION: FOUR THEORIES IN ONE

The word evolution does not accurately describe any of the four theories that must be addressed to explain life on the planet–The Big Bang, abiogenesis, speciation, or natural selection.

Under the umbrella of **Evolution**, there are two theories about origins: the origin of the universe and the origin of life. We also have two theories for replication: how the original life form spread, and how life as we know it today came to exist. Collectively those four theories are what people think is meant by the term "evolution." For any valid debate as to whether creation or evolution theory is true, all four theories must be addressed with satisfactory and plausible explanations for each and every one for it to be fairly compared to creationism.

If you talk about the origin of life and call it evolution, the evolutionist may sneer. "That's not evolution! That's abiogenesis, you fool!" The evolutionist is not being entirely honest but is technically correct. Abiogenesis is not evolution. It is only one of four component theories included as part of the "theory of evolution." But none of these four separate theories are really evolution, strictly speaking.

THE BIG BANG

The origin of the universe is best explained scientifically by the modified *Big Bang Theory*. The creationist theory of the origin of the universe is that God created it. These theories are not necessarily mutually exclusive.

ABIOGENESIS

The origin of life is addressed scientifically in the theory of *abiogenesis*. This theory attempts to explain how life spontaneously originated from inanimate matter. Biogenesis is a term, which means a living organism is produced by one or more other living organisms. It's a concept easy to understand and recognized as true. Life spawns new life.

Abiogenesis is a whole different matter entirely. The theory requires life to originate from non-life or inanimate matter. Interesting, the creationist theory of life explains very clearly how lifeless clay was shaped and became animated by the breath of life from God. The current scientific theory calls for lightning strikes or asteroid showers into some magic primordial goo and single-celled life forms resulted from the cosmic effect. No matter how implausible the theory might seem, it has to be true because the alternative is God and that is always more implausible.

SPECIATION

The third theory confused for evolution is the theory called *speciation*. Speciation describes how life could spread from the original, single-celled organism that resulted from abiogenesis. Through "descent through modification" this organism must account for every modern form of life on Earth today, both plant and animal.

Creationism contends God created the light, the earth, the seas, vegetation, living creatures in the water, birds, fish and fowl, land animals and finally man, in that specific order.

Interestingly, the order of creation in the Bible and the order scientists declare speciation follows are the same, but the Bible was written thousands of years before the development of theory of speciation. How did an ancient writer with no knowledge of the fossil record nail the order of creation perfectly? Was it a lucky guess?

NATURAL SELECTION

If you look at evolutionary theory as a whole, natural selection is the only theory that cannot be disputed, and there's a biblical reason to believe in natural selection, which is found in the story of Noah. [2] When God tells Noah to take two animals of every kind, He could not have meant two black bears, two grizzly bears, etc. because the dimensions the Ark weren't big enough for that many animals.

Darwin's theory of natural selection addresses the final problem of how mutations on a small scale led to modification on a larger scale that can only occur over longer periods of time than anyone could ever stay alive to verify.

Darwin's theory is the one most often used interchangeably as a term for evolution, which plays right into the evolutionist's hands because it's also the easiest of the four theories to prove. Natural selection is simply another term for genetic mutation.

Mendel studied mutation in plants and called natural selection in the plant world heredity. Darwin observed heredity in animals and called it natural selection, but it was more than simple genetic mutations. He also observed behavior originally coined "survival of the fittest," implying those animals with a physical or reproductive advantage have the best chance of survival.

WHEN SELECTION ISN'T NATURAL

The evolutionist cannot declare victory while abiogenesis and speciation are the only alternative theories to God's involvement in creation because those theories don't work. It would be nice to solve the problems we have regarding the source of matter without God for the Big Bang, too–if it were possible.

If belief in natural selection is weighted equally to belief in God, it isn't an "apples-to-apples" comparison of theory–it's a stacked deck. It's essentially comparing part of one theory against a full explanation.

Evidence can be cherry picked. Only the most convincing evidence of natural selection is pitted against evidence to support the theory of creation. Hard to believe and impossible to believe are two different things, and when the opportunity arises shortly, I will point to a couple of theories under the **Evolution** umbrella which are impossible to believe, quite frankly. And before you conclude there cannot possibly be any evidence for creation, please refrain until you've read the coming chapter titled Reason to Believe.

Heredity is another term for natural selection, which is merely a description for how genes alter individual characteristics from generation to generation, and how some characteristics become more frequent because they enjoy some survival or reproductive advantage.

We breed dogs using artificial selection; by crossing a Labrador Retriever with a Poodle, you have a new custom breed called a Labradoodle. Mom has blue eyes, dad's eyes are brown, and so the child will have brown eyes because brown is the dominant trait. The child won't look exactly like Dad because Mom's genes get mixed in too. Their kid is a miniature melting pot. By definition, offspring are always hybrid. Natural selection within the species boundary is normal, natural and believable.

Across the species boundary, it is none of those. It's difficult, but not impossible, to believe a horse could mate with a zebra. It would be very surprising if the resulting offspring was fertile if it lived at all. In fact, I'm not aware of any successful breeding across that species boundary that produced live fertile offspring, even in mating between similar species that do produce young that survive. A horse can mate with a donkey and produce a mule as offspring, but the mule is always sterile. In the nine months of research I've done for this book, I failed to find a single example of fertile offspring bred across the species boundary. But it would have to happen and happen rather frequently for speciation to really describe how life diversified. We'll save the issues that apply to speciation for further discussion a little later.

Evolution has proven difficult to define in a way that doesn't immediately distract debate from the argument of what evolution is to the tangential argument of what evolution is not. The term is thrown around as the solution to every problem, but scientifically speaking at the same time, the word means little.

Evolution today encompasses much more than a genetic progression from apes to humans or Darwin's theory of natural selection. The theory itself has "evolved" to include all four of the component theories required to explain life on earth: the Big Bang, abiogenesis, speciation, and natural selection. It's a catchall phrase; the perfect smoke screen for science to hide the fact that natural selection and the Big Bang theories are the two component theories of evolutionary belief are what most people accept as true and the rest of evolution theory has some very serious flaws.

It bothers me within the first few pages of many books supporting evolution and arguing against creationism that claims evolutionary theory is not theory but generally accepted scientific fact. Having concluded the task of defining evolution to be fraught with contradictions and complexity, I owe it to the reader to explain my definition of evolution and creation as I will subsequently use the terms for the remainder of this book.

PUZZLE PIECES

Evolution is The Big Bang, abiogenesis, speciation and natural selection combined into one comprehensive perspective of how the universe and life within it came to exist. An evolutionist is a person who accepts those four scientific theories as the sole explanation for life, the universe and everything.

Creationism is the theory that God created the universe, the heavens and the earth, water, life, and every individual species. The creationist believes the biblical account of creation describes the method by which God created everything.

Not all scientists believe the evolutionist theory in total and adapt their knowledge of science and their knowledge of the Bible to reconcile their differences of opinion. That is the approach I have taken in the development of my personal religious philosophy. I suppose this might also be the best time to point out that while I attend church and consider myself a member, I do not pretend my personal beliefs match church doctrine perfectly. But I never claimed to be perfect.

If you think of evolution as a kind of math problem where you're solving for X, in this case X has a four part answer. Chronologically addressing the relevant questions in order, we must first ask, "How did the universe come to exist?" This is where the evolutionist expectantly waits for the creationist to argue against the Big Bang Theory. I happen to believe both creation and the Big Bang Theory are most plausible.

The Big Bang requires accepting the premise the Earth is probably several billion years old and the Universe may be as much as 10 billion years old. The evolutionist is ready for this because he expects the creationist to say the earth can't be more than 6,000 years old. This is because of a common misunderstanding about the creation account in first chapter in the book of Genesis. If I believe the Bible, I'm not supposed to believe in an old Earth. But if you accept, for the sake of argument for now, the Big Bang simply explains *how* God created the universe, we can move on to the next step explaining life on earth.

And once you move beyond the question of the origin of the universe with evolution and creation theory more or less on the same page in respect to the Big Bang, the evolutionist immediately gets into deep trouble getting from there to natural selection.

The evolutionist's answer to the origin of life on a newly formed earth is the theory of abiogenesis, which Random House dictionary defines as "the now discredited theory that living organisms can arise spontaneously from inanimate matter." The important word in the previous sentence is "discredited." The theory of abiogenesis is a complete joke, but a joke that continues to be the only alternative to the breath of life from God so the evolutionist clings to a fallacious theory. There isn't a "Plan B," there's only the breath of life from God or abiogenesis as competing theories. Abiogenesis doesn't provide very much competition.

Assuming for a moment a single celled individual life form could somehow spontaneously come to exist, it must then survive. It must have adequate sources of food and water and recognize those as such. It would have to understand how to reproduce and realize the need and a reason to do so.

The need to reproduce must have been ingrained into whatever consciousness that original organism possessed. If you are the only living organism with abundant resources to yourself, reproduction is counter intuitive. Why create competition for yourself?

The Big Bang satisfies the answer to the origin of the universe provided God is the catalyst. Abiogenesis does not provide a satisfactory alternative to Intelligent Design.

Next is the question of speciation. From the original single celled organism, which must have been a plant so photosynthesis could provide a source of food, we have to have a mechanism that creates the different life forms we know exist today.

The evolutionist's argument is that speciation is a process very similar to natural selection (which we can observe occur within species) but takes much longer. This explains why we can't observe it today. This means every plant and animal in their respective kingdoms, every phylum, class, order, family, genus, and species must be related to the original single celled plant form. Every living organism on the planet resulted from this process called speciation. All that is required is enough time.

TROUBLESOME TIME FACTOR

There is a problem with the time factor of evolution. There have been at least five mass extinctions of life documented by the fossil record, followed by explosive growth and diversification in the life forms which rapidly emerged post-extinction. Because natural laws we observe today dictate that cross-species breeding is extraordinarily difficult if not impossible to produce fertile offspring, logically the fossil record contradicts and ultimately refutes the theory of speciation. The fossil record shows life didn't just come back and repeat the slow, gradual process we are told is a crucial element of speciation; life came back with a vengeance. Following each extinction was a veritable population explosion, followed by an almost immediate cessation of the process.

The creationist theory is that God is perfect and therefore every life form He ever created has been perfect. The point is moot. The more interesting question is this: if God created every creature, why have some been allowed to become extinct? Traditional creationist theory also fails to address the explosive growth and diversification of species after these multiple mass extinctions assuming God created every life form only once unless a day of creation was a God day and not a 24 hour earth day. But if God days of creation are correctly acknowledged to be much longer than a traditional Earth day as we know them, the fossil record evidence does not contradict creation theory at all.

Because I believe in extinction and creation, I had to envision a compromise theory between the evolutionist and the creationist. My theory is a vision of God as an artist of incomparable skill. During the six "days" of creation, could God have been prototyping creation in the early "days" and use those models to refine His design by the sixth and final day of creation? Some religious people believe God is rigid and inflexible because of His perfection, but the scriptures contain a number of examples where God changes His mind. One example is in the Book of Kings, found in the story of Hezekiah. God sent an angel to inform Hezekiah he was about to die. The king demonstrated great humility, wearing sackcloth and ashes as he mourned his coming demise. God was moved by Hezekiah's show of remorse to the point where He extended Hezekiah's life by 15 years.

The final theory of evolution is natural selection, which is essentially an alternate definition for change or mutation. Natural selection makes sense so there's no need to debate that aspect of evolution theory.

Apparently scientists who write books love numbered lists. When Berra ran out of things to say defending evolution after 122 pages, he filled the remainder of his books with facts about court cases and lists of information. One such list was titled "Some Creationist Claims - Do they raise some doubts?"

I would like to discuss a couple of points Berra made on his list.

First Berra mentioned transitional fossils such as *Archaeopteryx* which he says "is clearly intermediate between reptiles and birds" in order to imply speciation has proof. This statement simply isn't true. The words, which make the sentence incorrect are "is clearly." It appeared to the first zoologists who arrived in Australia that the duck-billed platypus was a living "missing link." It was originally thought to be a cross between a bird (duck) and mammal (otter). Observations proved the creature was simply a previously unknown kind of mammal.

The only thing clear about *Archaeopteryx* is that it is the fossilized remains of a dead bird that appears to have scales on its neck like a lizard. That much *is* clear. Since *Archaeopteryx* is extinct, there aren't any live examples to observe today and investigate.

The other problem I couldn't ignore was when Berra attempted to refute the assertion that evolution was also a religion and required faith by saying "Creationists are beginning to admit that their "science" is not science at all, and that it depends on faith, but, they are quick to add, so does evolution. Not so. Biologists don't have to *believe* that there are transitional fossils; we can examine them in hundreds of museums around the world, and we make new discoveries in the rocks all the time. Scientists do not have to *believe* that the solar system is 4.5 billion years old; we can test the age of the earth. We do not have to *believe* that protocells (amino acids bonded together as pre-cell building blocks) can be easily created from simple chemicals in the laboratory; we can repeat the experiments, with comparable results. We can also create artificial species of plants and animals by applying selection, and we can observe natural speciation in action. That is the big difference between science and religion. Science exists because of the evidence, whereas religion exists upon faith—and in the case of religious fundamentalism and creationism, in spite of the evidence." [10] [Emphasis original]

The lengthy quote above, from Berra, has several errors and misstatements. First, there are fossils that biologists *believe* to be transitional, not transitional fossils. The second fallacy is the suggestion we can accurately test the age of the earth to prove it is 4.5 billion years old. Scientists *believe* the earth is that old because of carbon dating, but carbon dating is only accurate up to 50,000 years old. The older the age of the artifact being dated, the more likely an error will be made estimating its age accurately.

Third, scientists must *believe* that creating a protocell in the lab is the same thing as artificially duplicating abiogenesis and it is not. Scientific experiments in laboratories have produced a chemical reaction that led to the formation of a couple amino acids that are not alive. It's like throwing a few pieces of wood and some bricks on the ground and calling it a house. Or I should say, a "proto-house?"

Finally, the claim science has observed natural speciation in the wild is completely false. Scientists observed the natural creation of a new breed of species in the wild, but two breeds of fruit fly that mate do not produce a new species, but a new breed of fruit fly. The gratuitous insult thrown in at the conclusion of his point about fundamentalist Christians believing creationism in spite of the evidence was both ironic and unnecessary.

I am perfectly willing to concede the Big Bang Theory is the most logical and probably the correct explanation for the origin of the universe provided God is recognized as its catalyst. I will stipulate the earth could be as much as ten billion years old and is probably 4-5 billion years old. Of course, I believe that the Big Bang only occurred when God spoke and said, "Let there be light." There is still a problem with the Big Bang even when you accept it, which is: from where did the original matter come?

I've been asked if my book is an attack on evolutionary theory or a defense of my religious belief. It's a little bit of both. I only point out the parts of evolutionary theory that make no sense. I have to defend my religious beliefs in the process, because the two issues are interrelated. The aspects of evolutionary theory, which I do believe are pieces that can fit into a puzzle of a Big Picture drawn by God. The parts of evolutionary theory I do not accept have not been rejected because God and evolution are mutually exclusive, but because I am being asked to believe something that makes no sense based on the evidence.

3 FRAMING THE DEBATE

There are currently two theories regarding the origin of life and the world we know as we know it. They are evolution theory and creation theory. Creation theory is the belief God created the heavens and the earth, and all life on earth. We just discussed what evolution theory really is. Only one of the two theories gets any serious study or discussion in educational forums these days and that, of course, is evolution. But what if you compare the two theories side by side and explore the strengths and weaknesses of each?

THE TRADITIONAL EVOLUTIONIST PERSPECTIVE

Evolutionary theory explains the creation of the universe, the origin of life, the spread and diversification of life, and the existence of life, as we know it today, as the following sequence of events:

Approximately ten billion years ago, the universe originated with The Big Bang, which occurred when all matter from which our modern universe is comprised was compressed into an incredibly small, super-heated mass that suddenly blew apart with such intensity that the matter continues to expand to this day. Before the Big Bang nothing existed–not even space. There was only this ball of densely compressed matter. With incredible precision, a number of cosmic variables known as Rees constants were perfectly tuned in order to allow the formation of stars and planets, the sun and moon and the earth. If the balance of mass were slightly increased or decreased, the ability of matter to coalesce and form certain elements would have been prevented. Had not Rees constants been perfectly aligned, planets would not have sufficient mass to form properly. But they were. After the Big Bang, the planets and solar systems organized themselves according to mass and gravitational force.

Roughly four billion years ago, the Earth formed. The first form of life originated about three and a half billion years ago, only a few hundred million years after the earth itself is believed to have formed. That process of self-initiated life known as abiogenesis occurred when several key amino acids joined together and with some form of external catalyst, perhaps a lightning strike or more likely comet debris struck the earth and caused a unique reaction that resulted in the first living organism. The first life form was in all probability a single-celled plant or bacteria of some kind.

Life itself cannot be artificially created in the lab—yet, but the formation of the primitive building blocks like essential amino acids can be duplicated. Amino acids form proteins, which is a key step in creating these building blocks mandatory for life to exist.

Given enough time, patience, and resources, the origin of life will be artificially proved in a laboratory one day. If not, it will only be due to the fact atmospheric conditions on the newly formed earth were very different than today's atmosphere. Therefore it is possible the prehistoric atmosphere cannot be reproduced to the degree abiogenesis will require. This lack of ideal atmosphere may explain why we have found no indication of life on other planets.

When life was established on earth via abiogenesis, there had to be a source of food and water available to sustain it. These first organisms were likely single-celled prokaryotes that relied on photosynthesis for food and had adequate fresh water. From this life form, all plants and animals eventually descended. This first organism could reproduce asexually and did so in order to perpetuate life.

This first plant was resilient enough to survive until it had enough time to gradually evolve into other life forms. Eventually, given enough time, this propagation led to the instantiation of every life form in the history of the earth progressing from single cells to more complex plant and then animal life. Animal life was initially believed to have begun in the oceans and moved onto land, but current theories include speculations that life began on land, moved to the water then back on land. Whatever the sequence, this process known as speciation resulted in the origin of all living things.

The resilient nature of the original life form was inherited by following generations. The quest to sustain itself led to the survival of a core population of organic life that managed to remain alive through at least five known mass extinctions. The extinction evidence in the fossil record and obvious evidence from dinosaur bones that massive creatures once roamed the earth demonstrate the weakness of the creation account as an accurate source of information.

Speciation accounts for the spread of life into all life forms, which have ever existed on earth. The development of every new species of plant or animal occurred in a process, which only occurs gradually over an extended period of time and cannot be observed. Speciation occurs over millions or billions of years rather than from generation to generation. The process simply takes longer than science can observe or replicate artificially via experiment.

Because we know of the truth in the Big Bang Theory and the truth of natural selection, we may safely extrapolate the truth of speciation will be proven in only a matter of time. Some proof already exists in the form of the well-known transitional fossil or "missing link" examples *Australopithecus*, *Archaeopteryx*, and the more recently discovered *Tiktaalit*.[11]

The process known as natural selection described by Charles Darwin in his masterwork *On the Origin of Species* helps explain speciation and proves it by observed examples. Evidence of speciation which occurred over a relatively short period of time was observed in studies of pepper moths and fruit flies respectively. The first study showed light and dark colored pepper moths were affected by natural selection. Natural predators and environmental factors such as pollution played a role in the physical characteristics of future moth generations. Likewise, two different species of fruit fly were observed to mate and breed a new species, providing multiple documented cases of speciation.

Darwin detailed his observations and coined the phrase natural selection. If you think of natural selection as evolution, it is micro-evolution and speciation is macro-evolution. Natural selection is true and obviously a fact, and Darwin documented how natural selection worked in detail. Because dinosaurs and other fossils in the geologic record are not documented in world religions but obviously exist, and given the fact that natural selection is an observable process that clearly occurs, one may reasonably conclude creation accounts are false and evolution is true.

THE TRADITIONAL CREATIONIST PERSPECTIVE

Creationist theory explains the creation of the universe as the product of a supernatural entity with powers far exceeding all capabilities of which the human mind can conceive and grasp. During the six days of creation, this supernatural God created everything we know as the creation accounts of multiple religions agree.

The Judeo/Christian/Islamic creation account is the literal, undisputed word of God. It is without error, scribed by authors inspired by God. The creation account is perfect as God is perfect. The creation account provides a clear narrative of how God created everything we know.

Originally there was nothing. God created light in the beginning, on the first day of creation. There were six total days or periods of creation, beginning with God's creation of the universe and the earth on the first day. On the second day, God created water and the sky and provided our atmosphere for the Earth. On the third day, God created dry land and vegetation. The sun, moon and stars were formed on the fourth day. The fifth period of creation saw the water "teem with living creatures, and let birds fly above the earth across the expanse of sky." God told the sea creatures to "be fruitful and increase in number to fill the water in the seas" and birds to "increase on the earth."[12]

All land creatures and man were created on the sixth and final day or period of creation. Biblical experts have added up all the ages of people listed chronologically from Adam and meticulously calculated the elapsed number of years from all generations that were specified in the rest of the Bible. These scholars advocate a young Earth theory, estimating the planet to be about 10,000 years old including the 144 hour long week of Creation plus the additional 24 hour day of rest.

Most people believe in a supernatural Creator who is omniscient, omnipresent, and omnipotent can literally do anything, including making the universe and everything in it in 144 total hours, so of course He could make the earth appear older than it really is. God can do anything. He is everywhere at all times. God knows everything at every moment. He knows what you are thinking at this very moment. His power is endless, perfect and complete.

While it is true the Bible does not specifically mention dinosaurs by name, there is an explanation for them that does not contradict biblical truth. There are also biblical references in the book of Job that are open to interpretation. The behemoth is said to feed on grass like an ox. My bible suggests in its footnotes that the animal meant in translation is an elephant or hippopotamus, but the description also says "His tail sways like a cedar."[13] That sounds more like a brachiosaurus than an elephant. Likewise, Job mentions the leviathan in the ocean, which could be a whale, or possibly an aquatic dinosaur.

But the Bible does mention dragons; perhaps these were dinosaurs. We know Piltdown man was a fraud and those other fossils must be as well, as Fred Hoyle pointed out about *Archaeopteryx*. Perhaps dinosaurs aren't as old as carbon dating indicates.

There is a logical explanation for the other transitional fossils…except we don't really need an explanation because creationists believe in God. It is unhealthy to ask too many questions about God. It is disrespectful.

DIVINE EVOLUTION: THE HYBRID THEORY

The above narratives provide a brief synopsis of the evolutionist and creationist perspectives from an impartial point of view. The strengths and weaknesses are described from the perspective of the theory advocate.

What if neither theory is right?

After carefully examining the arguments from both sides, I cannot fully embrace either. My own personal theory of creation is based on the same creation account information provided by the Torah, Bible, and the Qur'an, all in very similar accounts. The traditional creationist theory described creation as taken from the Old Testament of the Christian Bible. The Pentateuch or first five books of Hebrew Scriptures are the same five books known as Genesis, Exodus, Leviticus, Numbers, and Deuteronomy. The creation account in the Qur'an actually comes close to a Scriptural description of the Big Bang.

While I am not an Islamic scholar and to become one is well beyond the scope of this book, I performed some cursory research on the Internet to find a clear narrative of creation from the Islamic perspective. I found an article titled "Creation of the Universe" posted on About.com in the Islam section and corroborated by two different posts at IslamOnline.net, one by Nehal El-Hadi and another from Zaghlool El-Naggar, Ph.D. "Creation of the Universe" was written by a layperson named Huda who made the claim the Qur'an describes the Big Bang.

Huda wrote "When describing the creation of the "heavens and the earth," the Qur'an does not discount the theory of a "Big Bang" explosion at the start of it all. In fact, the Qur'an says that "the heavens and the earth were joined together as one unit, before We clove them asunder" (21:30). Following this big explosion, Allah "turned to the sky, and it had been (as) smoke. He said to it and to the earth: 'Come together, willingly or unwillingly.' They said: 'We come (together) in willing obedience"' (41:11). Thus the elements and what was to become the planets and stars began to cool, come together, and form into shape, following the natural laws that Allah established in the universe." [14]

It appears the traditional evolutionist looked at the origin of the universe and life and developed several theories, a couple of which still hold water. Traditional creationist theory has a couple of minor translation or interpretation issues. These are blown out of proportion when strict creationist "experts" insist on literal interpretations of every word in the account from their perspective. Strict creationists and strict evolutionists are remarkably similar in their lack of tolerance for any alternate point of view from their own.

I believe the Christian Bible is one of the most remarkable books ever written. Two people may read the same passage and interpret its meaning differently–and both could be right. The Bible is like a compass to direct us to the truth. When we pray for truth then seek answers for our questions in the Bible, we will find them. I know the Torah and Qur'an contain truths, but I must speak foremost of The Truth I have personally studied and accepted.

A most plausible hybrid theory of creationism and evolutionist beliefs is this:

God created the earth as the biblical account and evolutionists both describe the order of creation. Evolutionists fail to recognize the creation account follows the fossil record in the chronological order life appeared on earth. It only shows that creation happened in that order but over a much longer period of time than 144 hours. The young Earth time factor problem evolutionists use to condemn creationist theory is eliminated if one accepts the premise each God day of creation could be as much 100 million years in duration or more. The actual length of the day is the only sticking point.

When God said "Let there be light" matter itself was created, and with it the universe began with a Big Bang sparked by God. The creation account refers to the period of creation as a week of time. A day to God is certainly not a day of man, evidenced by the timing of the creation of the sun and moon. These heavenly bodies were not created in a timely fashion to be able to mark the passage of the first three days of creation.

There is no reason for a writer more than two thousand years ago to know anything about the fossil record. There is no reason for that writer to even know there WAS a fossil record. Yet the creation chronology explaining the order in which animals emerged virtually matches the order supported by the fossil and geologic records.

Most evolutionary biologists believe birds are descended from dinosaurs. Birds and sea creatures were the only animals created on the fifth day of creation. If each day of creation is actually millions of years long and dinosaurs really did mutate into birds, it all could have occurred within the confines of "Day Five."

God may have wiped the slate clean via extinction to start the sixth day and then created the modern version of each species. Each previous day's work was treated as a prototype or discontinued model (see Berra's car example) with few exceptions. Such enormous reptilian creatures as the dinosaurs may have been the equivalent of a divine "doodle in the margins" since they would have likely out-competed the species God eventually settled upon.

Douglas Adams had the brilliantly funny suggestion in his novel *The Hitchhiker's Guide to the Galaxy* that dinosaurs were never living creatures but actually models cast from molds which the fabricator of earth (not God, of course, but a friendly chap named Slartibartfast) embedded in the Earth's crust for his personal amusement and to make digging in the dirt more interesting for humans.[15] Of course that theory is as preposterous as speciation for the explanation, but at least it is funny.

You might have noticed an excess use of the words "may" or "could have" any number of times talking about these theories. Any time you see those words, it is implied the word "not" is also possible. For example, a day may not be longer than 24 hours even during creation. You will note no fewer or more instances of these words in divine evolution compared to creationist accounts or evolutionist accounts, because all are theory.

What you don't see are words like "absolutely could not" or "impossible," but when describing the flaws in evolutionist theory, one must confront impossible contradictions. It is impossible for something to come from nothing. Yet before the Big Bang there was literally nothing according to scientists (or at least my college professor), so where did the matter that exploded come from? Neither speciation nor abiogenesis can be proven in the lab, so why is either still considered scientific theory? Why is meat from a pig wrapped in a sausage like skin called a hot dog?

Ok, that last one wasn't evolutionist theory…just seeing if you were paying attention.

One cannot honestly state with absolute conviction any fact about which any doubts remain. Dawkins describes a scale ranging from 1 to 7 where 1 is a pure Christian (no doubts) and 7 is a pure atheist, and even he won't give himself more than a 6. In contrast, I would give myself no less than a 2, because as strong as my convictions are, I could not swear in court to what I believe to be true, but only what I know is true. For some reason, Dawkins expects creationists to all be 1's but the most avid atheists can't muster more than 6.

Evolutionists try to link disparate species by noting similarities in their genetic makeup. For example, the theory of speciation's progression of ape to man was changed to range from chimp to man after it was noted chimps and humans are genetically more closely related than ape and man.

Rather than speciation resulting from mating between two different species as some modern scientists have claimed to have observed, the ancient form of speciation allegedly occurred when an ancestor species "forked" at some point in their evolutionary development to form two disparate species. In other words, when we see evidence of speciation in action today it is of a different type than what worked in the past.

In his book *Next*, author Michael Crichton wrote, "When the human genome was decoded, scientists were startled to find that it contained only about thirty-five thousand genes. They had expected more. After all, a lowly earthworm had twenty thousand genes. That meant the difference in complexity between a human being and a worm might only be fifteen thousand genes. How then, could you explain the huge difference in complexity between the two?" [16]

Crichton pointed out that humans might have less than twice the number of genes in their genetic makeup when compared to the lowly earthworm, but later he also points out that it is the combination of those genes that make the complexity of life so special. Not every gene can be combined or paired with just any other.

While I'm not in agreement with every word Crichton wrote, it isn't because I doubt the veracity of his research or diligence he put forth. I only take issue with his assertion there are "regions of meaningless code" in genes, preferring to believe they are instead regions of code of which we do not yet know the meaning.

The fact that ancient scriptural writers who didn't study fossil records could get creation in the right order is too convenient to be coincidence. There is no claim by any creation account that says the story is comprehensive in every minute detail. God didn't iterate every species He created. The creation account covers every milestone indicating which animals came in what order. The creation account agrees with the fossil record on that order but furthermore it gives a general description of *how* God created life.

The ancestral form of forked speciation does not explain how life expanded and diversified after each of the mass extinctions indicated in the geologic record. Evolutionists have crafted a theory called punctuated equilibrium to attempt to solve the problem of the multiple mass extinctions clearly indicated by the fossil record that undercuts speciation theory's requirement for long periods of time to get things done.

Punctuated equilibrium describes speciation as a process, which occurs in spurts when it is necessary to account for the rather sporadic explosions of life. These explosions are followed by long periods where no new life forms emerge. For whatever reason, this process remains dormant for eons until it is needed. Then it is modified apparently in order to compete with belief in a Creator God because every other plausible or rational explanation failed to work.

Explaining the mass extinctions and ancient, buried relics like dinosaur bones to the writer of the first books in the Bible was unnecessary. What relevance would a creature dead for millions of years be to the ancient Israelites? It's called "need to know." For that matter, other than curiosity, why do we need to know? What do dinosaurs tell us? Unless you aspire to a career in paleontology, the usefulness of this information will end after the final exam for your science class.

Dinosaur skeletons are awe inspiring to view if you visit the Smithsonian museum in Washington. They are fascinating and entertaining–but do they provide useful information? It depends on your line of work. If I found myself thrust into a real life *Jurassic Park* and dinosaurs again roamed the earth, knowledge of their eating habits and running speeds might be relevant. Otherwise the net benefit from knowing intimate details about dinosaur life has at best been good for a correct answer or two in *Trivial Pursuit.*

Current theories of abiogenesis and speciation are both ridiculous and impossible. Both are impossible to prove and are purely speculative theories only given credence because the Big Bang and natural selection serve as scientifically proven bookends. Abiogenesis occurred when God breathed life into inanimate matter. Speciation occurred when God shaped that same soil consisting of amino acids and proteins and formed birds or reptiles and every other life form and breathed life to animate what was not living. If each day of creation spanned about a billion years, the creationist perspective becomes much stronger and more plausible and the evolutionist theory is conversely weakened because it draws strength from denigrating creationism.

Speaking of impossibilities, the Big Bang was impossible without God either creating or bringing matter to the event. If matter was not the light created when God said, "Let there be light," the light must have resulted from the release of massive energy as matter.

In the traditional creationist account, only a single word is misunderstood, and that is the definition of what a day means to God. An Intelligent Designer or Supreme Being is not confined by human definitions. His power and the magnificence of His creation are not diminished by spanning a period of time greater than 144 hours.

Interpreting the creation account definition of a day differently makes sense because the account describes God creating the sun and moon on the fourth day of creation. These are the instruments by which humans measure the length of a day. Without them, a day could be any length of time.

Is there any biblical evidence to substantiate the claim the first six days of creation were longer than typical Earth days? As a matter of fact, there is. In Psalms 90:4, David wrote "For a thousand years in your sight are like a day that has just gone by." Paul echoed David's words in II Peter 3:8, which reads "But do not forget this one thing, dear friends: With the Lord a day is like a thousand years, and a thousand years are like a day."

Arguing a negative does not prove a positive. If the negative argument fails or is discredited there is the risk of a backlash effect that not only neutralizes the attacking argument, it strengthens the point previously under assault. When people argued evolution was false but evidence showed natural selection was true, it damaged the traditional creationist perspective and bolstered evolutionist theory. I must be careful to remember to acknowledge that parts of evolutionist theory appear to be reasonable and true.

If solid evidence of speciation or abiogenesis is one day presented, the evolutionist theory may be revisited. The new information may be applied to either buttress or refute Divine Evolution. Much of creationist theory appears to be reasonable and true, with the sticking point which remains to be the definition of a day. Divine Evolution explains how creation occurred by fixing the confusion over the meaning of that one word, bolstering the case for a modified form of creationist theory we now call Divine Evolution.

Now that the framework of thought has been established, the creationist theory I will argue from this point forward is Divine Evolution. Traditional evolutionist theory has value but is flawed. Traditional creationist theory has value but is flawed. Creationist flaws are minor compared to problems with evolutionist theories, but to deny any flaws exist is disingenuous.

Calling the creationist theory flawed does not make God flawed. It makes literal interpretations of God's work potentially flawed. The Bible was written by human writers who were inspired by God. I happen to believe writing this book is part of my purpose in life and was divinely inspired by God.

But I am not flawless and neither is this book. I'm not comparing my book to the Bible in any respect other than to assert my inspiration for this book came from a divine source and only in that respect are they similar. From that perspective, it is theoretically possible, but not very likely, the author of Genesis couldn't comprehend the God day concept and substituted a meaning he did understand.

In religious Scriptures full of analogies and parables, this synonymous use of God day and Earth day seem quite plausible. The point is, don't fixate on one detail to the exclusion of a wealth of evidence supporting a given theory. Always look at the big picture.

My personal view is the account was written purposefully as directed by God in order to allow for re-interpretation of the Genesis story from the ancient to medieval and now modern times. Each reading of the words fit the context appropriate for that reader's knowledge and sophistication, resulting in quite legitimately different but accurate interpretations of the same scriptural text. The understanding each reader derives is determined largely by what is cogent to the dominant civilized culture of the time. God speaks across the limitations and boundaries of time and language.

After reconciling the differences between the traditional evolutionist and traditional creationist views in this hybrid theory, my faith in God remains constant. The focus of the debate now centers on the possible motivation for the inevitable hostility of pseudo-scientists toward such a compromise theory.

Such hostility may exist due to the atheist's misunderstanding of God, but why becomes irrelevant if one chooses to accept the idea of Divine Evolution. Acknowledging God might really exist legitimately and completely shifts the debate from the scientific to the psychological.

Dawkins described religious people as delusional because they believe in something he doesn't, but he believes in impossible things of their own right. Are scientific people delusional to think every question we might consider about our world and this life has a logical and rational explanation, implying God is neither? Science cannot explain certain phenomena we know to be true. The only choice the strict evolutionist has is to ignore those phenomena and pretend they do not exist.

In summary, Divine Evolution is not a scientific theory but a philosophical theory formed by evaluating physical scientific evidence as well as psychological, spiritual, and even paranormal sources of information in search for truth. It was research in a quest for knowledge formed by (unintentionally) following the scientific method. A question was asked, research was done, a hypothesis was formed, observations were made, conclusions were drawn, and this approximates the report of my findings.

The movie *The Usual Suspects* ended with one of the great lines in the history of cinema: "The greatest trick the devil ever pulled was convincing the world that he didn't exist."

The correlative argument to the idea the devil does not exist is that God doesn't exist, either. But we know evil is real, and man commits atrocities every day in spite of his race, creed, or the color of his skin. Is evil spawned by Satan or inherently in man's nature?

Evil is evil–but is it inspired by evil, or innately present? If it's just there, why aren't we all equally despicable? Why do some of us strive to do good more than others? That is the real crux of the debate…

4 THE CRUX OF THE DEBATE

The idea for this book originated because I witnessed an unchallenged remark so ludicrous on its face that it begged to be addressed–if not then, now. The idea that man is intelligent because he can make a few interesting artifacts is debatable. To deny the designs he cleverly copies and raw materials he uses were borrowed from an intelligent creator God is an insult to both God and my intelligence. But the point of Dawkins calling his book *The God Delusion* was to insult me, was it not?

Dawkins was correct in pointing out the logical flaw of believing in a literal period of creation lasting only 144 hours. Whether it was possible to create everything in so short a time span is irrelevant. When framing the debate pitting traditional evolutionist theory versus traditional creationist theory, we found major flaws in pure evolutionist theory and between strict Young Earth Creationism interpretation of the creation account and scientific evidence regarding the age of the earth. The only reasonable alternative at that point was to develop my theory called Divine Evolution, stating God's six days of creation were actually millions or even billions of years long. However, it was during those creation periods every plant and animal, every species of creature and every living organism that we know today were designed by God.

ONE SIDED ARGUMENT

My greatest concern when I began this effort was due to something I read in *The God Delusion*. Dawkins said he would not debate a creationist because of his opponent's lack of scientific background. But it also occurred to me that Dawkins is a zoologist and not a psychiatrist yet feels qualified to judge me delusional without ever meeting me.

I guess qualifications are subjective in nature. Dawkins gleefully quoted a colleague who said when asked about debating creationists "That would look good on your CV, not so good on mine." I confess I didn't immediately get the insult; I had to translate CV to curriculum vitae and from that to American English to realize he questioned my credentials.

It is true that my bachelor's degree was in business specializing in computer science, and from there I spent twenty years in the field as a computer scientist writing software applications for big business. So my qualifications as a scientist to discuss abstract philosophical theory and religious issues are just as impeccable as Mr. Dawkins, in my opinion. That's my story, and I'm sticking to it.

It is interesting the same zoologist who insists we look at ancient fossils of long dead creatures and observe the "obvious" connections that appear to link two modern species to animal remains embedded in rock can also see overwhelming evidence in creation as only an illusion of intelligent design. Dawkins calls the religious believer delusional for accepting the most obvious explanation and the most simplistic.

While the simple solution is not always true, the most convoluted solution is never elegant and rarely viable. Insisting on a strict evolutionist approach to the available scientific evidence and ignoring the possibility of God is stubborn and somewhat indicative of an innate inability to think clearly. Dawkins goes so far as to fabricate a theory called memes to attempt a non-religious explanation for why people continue to believe in God generation after generation. Memes are like genes in that they replicate behavior rather than physical characteristics, but unlike genes because they are invisible, they don't really exist. They do exist in the mind of Richard Dawkins and a few of his acolytes who churned out any number of books perpetuating this nonsense, the most famous of which being Susan Blackmore's *The Meme Machine*. But not all books which talk about memes agree with Dawkins' theory. The last three essays in the book *Darwinizing Culture* by Robert Aunger effectively refute the theory of memes and expose it for the pseudo-science that it is in devastating fashion. Notwithstanding the fact, any number of truisms may be applicable such as the KISS principle (short for Keep It Simple, Stupid) or the cliché "the shortest distance between two points is a straight line" exactly, because they are generally true statements; it is also true that as a computer scientist I learned the simplest solution doesn't always solve the problem. The true rule of thumb is the best solution is the one that is the most obvious, which solves the problem with the least amount of complexity possible, in order to minimize error. Keep it simple AND solve the problem.

Elegant solutions are never excessively complicated, and discarding the most obvious solution without good cause is just plain silly. The traditional evolutionist has a problem. He mistook young Earth creationist theory as the only alternative to traditional evolution theory and given an either/or choice, he chose the theory with the most scientific evidence available to support it.

However, divine evolution eliminates the problems with both traditional evolutionist theory and traditional creationist theory. So where will that leave us? In the process of defining evolution theory, we exposed the missing catalyst behind the Big Bang and the complete lack of substance to abiogenesis and speciation theories. In short, there was ample reason to reject evolutionist theory on face value and insist on dealing with the individual component theories. Combining the best of both worlds, placing scientific evidence and facts into creationist context and resulting in a seamless, much stronger and more logical hybrid theory is the simple, honest, common sense solution.

The obvious merit in evidence for the Big Bang and natural selection warrant further discussion of evolutionist and creationist theory, as well as divine evolution and any other reasonable and plausible explanations that may exist or be developed in the future. Evolutionist theory should not be taught in school as established fact. Teaching theory as fact is indoctrination. Teaching demonstrably false theory as fact is dishonest as well.

THE RELUCTANT EVOLUTIONIST

One of my favorite authors was Michael Crichton, whose intellectual brilliance seems underappreciated when compared to his popularity as a writer. Not every fan of his works realizes he was educated at Harvard intending to become a doctor and created the television series *ER* based on experience; he rewrote the classic tale *Beowulf* as a challenge, called his work *Eaters of the Dead* and it was made into a movie that starred Antonio Banderas titled *The Thirteenth Warrior*; his first novel, *The Andromeda Strain*, was written while he was still in medical school and became an instant sensation. His novel *State of Fear* exposed the fraudulent science behind global warming years before the Climategate scandal came to light, and *Next* speculated about the potential horrors to come that will result from increased breakthroughs in genetic research from legal and physical struggles for right of ownership of one's personal body tissue.

Michael Crichton wrote *Jurassic Park*, a novel about abuses in scientific advancement via technology by rash and ruthless businessmen seeking obscene profits. The greed of certain people allowed the release of genetically re-engineered live dinosaurs to roam a remote tropical island. In the process, the dinosaurs ate a few unsuspecting visitors to the island. The tremendous success of the book and the corresponding major motion picture grossed over $900 million dollars worldwide and created a franchise in the process. Crichton wrote a sequel novel called *The Lost World*. Somehow Hollywood managed to get two movies from one sequel, releasing *Jurassic Park II* in 1997 and *Jurassic Park III* in 2001.

In *The Lost World*, Crichton spent an entire chapter titled "Problems with Evolution" exposing the real problems of evolutionist theory. Crichton begins the chapter by sketching out a brief history of evolutionary theory beginning with Baron Georges Cuvier up to contemporary theory. Cuvier was the first to realize old dinosaur bones being dug up were from extinct animals.

Crichton noted Cuvier "reluctantly" believed in extinction, but not evolution. Crichton also noted Darwin was the first to claim animals evolved. Using the character Ian Malcolm as his speaker, Crichton said: "Darwin amassed a tremendous amount of factual data—he had made an overwhelming case. So gradually his idea of evolution was accepted by science and the world at large. But the question remained: how does evolution happen? For that, Darwin doesn't give a good answer." [17]

Darwin's solution or natural selection is offered, which Malcolm accepts as a definition, but not an explanation. Using Malcolm as his voice, Crichton pointed out Darwin had no idea how evolution takes place. Malcolm succinctly sums the issue up by saying, "In the end, evolution is just the result of a bunch of mutations that either survive or die."

So far, no permanent damage has been done to the evolutionist argument, but Dawkins would have to be a little disconcerted by the direction Crichton appears to be taking. But immediately it gets worse, as Malcolm attacks abiogenesis, pointing to the time problem.

"A single bacterium–the earliest form of life–has two thousand enzymes. Scientists have estimated how long it would take to randomly assemble those enzymes from a primordial soup. Estimates run from forty billion years to one hundred billion years. But the earth is only four billion years old. So, chance alone seems too slow." [18]

Crichton (as Malcolm) then reminds his audience that bacteria actually appeared on the earth only 400 million years after the earth was formed, bringing up the theory Dawkins has embraced, which is an extraterrestrial catalyst for abiogenesis explains the time problem. Dawkins even goes a step further and says the "spaceman" theory is much more probable than a creator God. But wouldn't a creator God also be extraterrestrial by definition?

Next Malcolm eviscerates the evolutionist's argument against irreducible complexity, the theory by creationists that says certain animals display unmistakable characteristics of design based on the intricate details regarding their physical nature, man being the ultimate example.

Irreducible complexity reminded me of Dawkins' account of a court trial where a man named Behe used a mousetrap analogy to explain his theory of irreducible complexity. The memory made me laugh as I also recalled the lawyer I watched on You-Tube as he attempted to demonstrate his argument that "once and for all" destroyed the case for irreducible complexity with a live demonstration.

This lawyer removed the catch and the pin from a rat trap and proceeded to explain the irreducible complexity argument had been destroyed because the trap was still useful even with its reduced form. He demonstrated the trap still had functionality by attaching it as a tie clip, albeit the most ridiculous tie clip ever conceived and hardly functional. This lawyer further speculated it would be the ultimate weapon of war in a spitball fight or make an excellent paper holder. The harder the man tried to make his argument appear serious, the more laughable it became. The words "grasping at straws" came to mind.

But back to Crichton (again as Malcolm),who points to the "coordination problem" of evolution and uses bats and an example.

"Yet when we look closely at animals, it appears as if many elements must have evolved simultaneously. Take bats, which have echolocation—they navigate by sound. To do that, many things must evolve. Bats need a special apparatus to make sounds, they need specialized ears to hear echoes, they need specialized brains to interpret the sounds, and they need specialized bodies to dive and swoop and catch insects. If all these things don't evolve simultaneously, there's no advantage. And imagining all these things happening by chance is like imagining a tornado can hit a junkyard and assemble the parts into a working 747 airplane. It's very hard to believe." [19]

Malcolm (or Crichton, if you prefer) seems to agree with Fred Hoyle's famous tornado in the junkyard analogy, but that analogy is not strong enough in my opinion. For a bat or a human to exist without design involved is as likely as a tornado passing through a junkyard producing a fully functional 747, but given the fact biogenesis (mating and reproduction) is required, the tornado would need to produce both a male and a female plane in order for offspring to be possible.

The next problem Malcolm tackles is that evolution does not always appear driven by a "blind force" pointing out that some plants don't get eaten and certain animals haven't evolved as the rest of their contemporaries must have done. For example, he points out that sharks haven't changed significantly for more than one hundred and sixty million years and opossums haven't changed much since dinosaurs became extinct, sixty-five million years ago—in spite of dramatic changes in the environment.

After he thoroughly eviscerated the theory of evolution, Malcolm was posed the big question: "Are you saying evolution is directed?" but he responds, "No, that's Creationism, and it is wrong. Just plain wrong." [20]

And there the discussion of evolution ends between Crichton's characters in a manner remarkably similar to how evolutionists such as Richard Dawkins approach the debate with creationist theory. Dismiss it and move on. Don't bother with an explanation, or worry about not having one.

Crichton tantalized us by completely eviscerating evolutionist theory using known facts about evolution, but then he completely dismissed creationist theory in total without any discussion or even consideration. Unfortunately, Dr. Crichton has passed away so I cannot offer Divine Evolution theory for his consideration.

He got right to the brink of the debate on whether or not a Creator is a viable alternative, then backed off without explanation. Perhaps it was to maintain his mass popularity, but it is entirely possible that Crichton didn't believe in either creation theory or evolution theory. He says creationism is "just plain wrong," but he doesn't say why. He's right to acknowledge it is very difficult to believe life appears to have design but really does not.

Crichton sets up the opportunity for Divine Evolution but fails to offer it or any other alternative. He only throws away both creationist and evolutionist theory without replacing them with a viable alternative. But he wasn't completely finished with the topic yet.

Malcolm continued "...extinction has always been a great mystery. It's happened five times on this planet, and not always because of an asteroid. Everyone's interested in the Cretaceous die-out that killed all the dinosaurs, but there were die-outs at the end of the Jurassic and Triassic as well. They were severe, but nothing compared to the Permian extinction, which killed eighty percent of all life on the planet, on the seas and on the land.

No one knows why that catastrophe happened. But I wonder if we are the cause of the next one...maybe that's our function. Maybe every few eons, some animal comes along that kills off the rest of the world, clears the decks, and lets evolution proceed to its next phase." [21]

Or perhaps God initiated the extinctions a few times during the six God days of creation.

Crichton used Malcolm to decimate speciation theory, referring to the theory of punctuated equilibrium but giving no reason to believe it. Crichton offers no explanation as to how or why punctuated equilibrium could have happened, only noting the extinctions occurred, and that created the problem with the fossil record regarding speciation.

Malcolm doesn't appear to really believe in evolution, but he also knows that he can't prove a competing theory is true so he doesn't bother trying. His intelligence has led to developing so much pride and arrogance he can no longer believe any kind of superior being greater than him. If Malcolm was a real person, he might get along with Richard Dawkins fairly well.

Another character named Thorne astutely commented to an observer, "Are you listening to all that? I wouldn't take any of it too seriously. It's just theories. Human beings can't help making them, but the fact is that theories are just fantasies. And they change. When America was a new country people believed in something called phlogiston." [22]

He proceeded to point out phlogiston was a disproved theory once firmly believed, and also noted that people once believed that four humors controlled behavior and the earth was only a few thousand years old. When dinosaur bones were discovered, people finally realized the earth must be older than originally thought.

Then Thorne observed, "Now we believe the earth is four billion years old, and we believe in photons and electrons, and we think human behavior is controlled by things like ego and self-esteem. We think those beliefs are more scientific and better…They're still just fantasies. Have you ever seen a self-esteem? Can you bring me one on a plate? "[23]

The character of Thorne may not be portrayed to be as intelligent as Malcolm, but the wisdom of Michael Crichton becomes obvious no matter the "speaker."

Thorne concluded his common sense observations of the world by asking, "Do you feel the way the boat moves? That's the sea. That's real. You smell the salt in the air? You feel the sunlight on your skin? That's all real. You see all of us together? That's real. Life is wonderful. It's a gift to be alive, to see the sun and breathe the air."

How true. Crichton reminds us the essence of what is really important—reality, not theory. Theories about evolution, the origin of life and the origins of man are really just talk and conjecture. They have limited value. It can be entertaining to debate these topics ad nauseum. Thorne put evolution, what happened to them on the island into the bigger picture and then gave it proper perspective for the child. Indeed, life is a precious gift to be enjoyed each day. But if life is a gift, should one not logically conclude there was a Giver?

GENETICS AND LEGO THEORY

I developed computer applications for over twenty years. When I first began writing code, computer applications were monolithic blocks of code. Applications could not share technology. Each application that wanted to prompt a user to enter data into a field had to write code to perform that task. As a result, the appearance and complexity of each program varied widely and each program started from scratch. Eventually application developers began to recognize common functionality and coded routines to perform work as generically as possible in order to make the code reusable.

The advent of Microsoft's Windows operating system changed the dynamic of computer programming and the types and structure of the applications subsequently developed. A new means of grouping information into a module called a dynamic link library (DLL) was invented which allowed applications to organize code and merely call a method in this library of common functionality.

Once the importance and power of separating code into common blocks was realized, the antiquated monolithic applications were destined to become obsolete. The new technological breakthrough was in programming languages, where the language C++ allowed code to be grouped as objects that formed building blocks.

For example, an abstract object called "Printer" could generically allow a developer to call a Print method, and the runtime code could instantiate a derived Printer object of type OfficeJet7410 to handle the specific interfaces to a Hewlett Packard Office Jet, which has scanning and fax capabilities, or a

LaserJet6 object to handle straight printing for the printer named HP Laser Jet 6.

I used to work for Hewlett Packard, and I still love their printers, in case you're wondering about my examples. As a result of object-oriented programming, developers learned to write small, efficient and reusable blocks of code, and then build more complex applications inheriting these common blocks and adding functionality.

Occasionally during my professional career as a computer scientist, I was asked to teach a course about some technology I had helped develop. I was perceived as more conversational than the average developer and, therefore, was frequently drafted for duty as customer liaison or course instructor. As an instructor attempting to convey C++ theory to novice developers, I would invariably rely on an analogy between C++ objects and the children's toys known as Legos.

Legos are the small, rectangular plastic blocks that snap together and may be used to make complex toys by following some design or instructions provided. The blocks may be used to construct a pirate ship, a castle, or a rocket headed for space exploration. The limits of what may be constructed using Legos are set by one's imagination.

Scientists have noted the slight difference in genetic makeup between man and chimp, and that as few as fifteen thousand genes separate man from the worm. But they also learned the complex interaction of genes. Legos will connect to each other in the sense one block can be snapped into the connectors of another, but to achieve the design, one must attach the right blocks in the right places.

Genes are similar to Legos in the sense they "snap" together to form chromosomes, but the rules for connecting genes are far more complex. Not just any two genes can attach together.

We have barely scratched the surface of studying genetic behavior, and just know enough to be dangerous. The fact there are rules that apply to genetics implies design on the micro-level, which is order. Order is evidence of intelligence, opposite of chaos, which is the natural state of being.

WHAT DO TRANSITIONAL FOSSILS MEAN?

Fossils described as transitional are those, which appear to be a common ancestor for more than one species. Transitional fossils are, invariably, perceived to be intermediate forms or "missing links" between two separate species. Actually, these alleged transitional fossils bridge above genus or species level and theoretically occur as high as the class or order level, according to rules used to classify living organisms in the Animal Kingdom.

These fossils are viewed as critically important proof of speciation in the evolution theory, and seen as the only hard evidence for evolution apart from observations made about natural selection. The problem lies in the fact these fossils are only obviously dead and extinct animals and nothing more definite than that. The suggestion these fossils provide conclusive evidence of speciation is extremely questionable, for four reasons.

First, consider the duck-billed platypus. Were the creature extinct, it would be the most definite proof in existence in terms of a transitional fossil. The problem is when zoologists initially observed the live creature in its natural habitat they thought they were observing a hoax. The platypus has the bill of a duck, the body of an otter, the tail of a beaver, and venom like a poisonous reptile. However, it does not gain a unique classification and has not been declared a missing link. It is simply viewed as a weird mammal, not transitional. Why? Because the platypus is not extinct, and scientists have been able to study live specimens and learn from them. If a species can only be considered transitional if it becomes extinct, this further complicates speciation.

Second is the fact the type of speciation evolutionists claim must have occurred to create a transitional fossil is a kind of forked evolution. A distant ancestor mutates into two separate and distinctly different animals, possibly even of different classifications or orders. Modern speciation that is claimed to be observed by evolutionists involves two "species" of fruit flies or pepper moths forming a new "species" via natural selection. The latter form of speciation is not speciation at all, but evidence of natural selection naturally producing breeds similar to how we humans artificially create breeds of dogs. The fact neither kind of speciation can be reproduced via scientific methods completely invalidates the theory of speciation.

Third, the fail safe argument for the evolutionist is to point to common traits noted in the genetic makeup of similar species to claim DNA similarities between divergent species indicate these fossils are the "missing links" they appear to be.

In other words, we should believe our eyes when it is convenient and it supports evidence of speciation, but disregard the overwhelming appearance of intelligent design when we consider creationist theory. Allowing for a moment the hypocrisy as the evolutionist attempts to have it both ways, let us remember that looking at DNA or genetic evidence as some sort of proof of evolution requires one to ignore other possible explanations that make as much or more sense.

Fourth, although scientists acknowledge that fallacies in the fossil record would invalidate speciation theory, the typical response to a problem with the fossil record is to excuse the problem as an anomaly. For example, instances where older fossils are found above newer fossils by scientific age estimates, geologic faulting is blamed for shifting rock strata and creating the false impression by forcing one layer of rock over another.

There is also the coelacanth (or "dino-fish") believed to date back as far as 400 million years and thought extinct for 65 million years until a live specimen was found on a fishing boat in 1938 by a South African museum curator named Marjorie Courtenay Latimer. Since then dozens of coelacanths have been found living near the Comoro islands and in Indonesian waters. The status of the coelacanth has been upgraded from extinct to endangered, by necessity. Berra refers to the coelacanth as a "living fossil," apparently failing to recognize the oxymoronic nature of the phrase.

Berra simply reinterprets the significance of this violation of the fossil record, asserting the fish "manifests the phenomenon of arrested evolution" [24], whatever that is supposed to mean. What it really means is pseudo-scientists guessed the fish were extinct, and now we know they are not, so the template must be revised without changing the underlying system of beliefs. When asked by an exasperated skeptic what proof in the fossil record would allow an evolutionist to revise his theory regarding the meaning of fossils, one sarcastic response was "Fossil rabbits in the Pre-Cambrian."

When I began my research, I was aware of two major "missing link" fossils known as *Australopithecus* and *Archaeopteryx*, which are typically treated as conclusive proof of speciation. I ran across another major fossil discovery called *Tiktaalit*, purported to have been discovered more recently. This only meant there were now three major so-called transitional fossils rather than two. But even thirty or three hundred of these kinds of fossils don't prove speciation. The fact that speciation cannot be caused by experiment to produce fertile offspring above the species boundary belies the validity of speciation theory.

Assuming the evolutionist/creationist theories are either/or in nature is to assume one or the other is true and correct and the other is not, therefore no middle ground exists. Divine Evolution includes the most viable explanation for intelligent design by applying scientific evidence, only excluding aspects of each theory that fail the common sense test. Yes, there are similarities in genetic structure in every living organism. If every living organism shares as a common denominator a Creator to design their existence, would not common sense suggest the Creator did not "reinvent the wheel" for every form or iteration of his creation?

After designing such a magnificent building block for life like the gene, why not make it reusable? The gene, which builds the chromosome pairs in every living cell and ultimately defines the appearance of every living organism on the planet, is simple by itself. The complexity occurs from the billions of interactions these genes can exhibit, but that is not to say any two genes can connect to each other.

Of course, examination of the fossil record does lead to consideration of several interesting issues regarding creationist theory. As noted repeatedly, these issues usually are due to the six "earth" days of Creation translation.

Within the framework of Divine Evolution's postulate the God days of creation were substantially longer than 24 hours each, knowing that dinosaurs could have easily come and gone before the final day of creation where modern creatures took their form, could the dinosaurs have actually evolved into birds as some scientists have speculated? As far as the creationist theory assertion that God only created every creature in one form one time, is that because for God to be perfect He cannot ever waver or change His mind? Scripture tells us we should know better. God sent Isaiah to King Hezekiah to inform the king of his impending death. When Hezekiah donned sackcloth and ashes to show his humility before God, Yahweh was so moved by Hezekiah's display He sent Isaiah back to tell Hezekiah fifteen years would be added to his life.

Consider the theory of God as the master artist and inventor who prototyped creation. Were the dinosaurs possibly models for future life to be crafted from improved future design? Why is God considered to be diminished if He improves on an idea?

Today's science textbooks still offer a modified version of the "missing link" theory, claiming: "Thus, humans are not descended from chimpanzees or from any other ape living today but from a species that no longer exists." [25]

The authors of *Science, Evolution and Creationism* wrote, "If the common ancestor of two species lived relatively recently, those two species are likely to have more physical features and behaviors in common than two species with a more distant common ancestor."

The authors also helpfully point out, "Humans are thus far more similar to chimps than they are to fish" and go on to state "the common ancestor of humans and chimpanzees was a species estimated to have lived between 6 to 7 million years ago, whereas the common ancestor of humans and the puffer fish was an ancient fish that lived in Earth's oceans more than 400 million years ago."

But the five mass extinctions cloud the logic of that last assertion.

Yet the authors continue, "...based on accumulating fossil and molecular evidence, the common ancestor of humans, cows, whales, and bats was likely a small mammal that lived about 100 million years ago...A person writes, a cow walks, a whale swims, and a bat flies with structures built of bones that are different in detail but similar in general structure and relation to each other."

It is true that there are similarities in structure and relation between all living creatures and DNA is that common denominator. But what does it mean to say that all life came from the same building blocks? The authors of *Science, Evolution and Creationism* continue their train of thought: "By comparing the DNA sequences of two organisms, biologists can uncover the genetic changes that have occurred since those organisms shared a common ancestor." [26]

In addition to mass extinction evidence collected, which indicates at least five major catastrophic events wiped out most life on earth have occurred within the last 450 million years and makes the speciation theory much less believable, there are other problems such as inconsistencies in what we have observed about the fossil record.

For example, horses started as small animals and evolved into larger ones. Elephants descended from mammoths considerably larger than the modern day model. The only reason we don't perceive these incarnations as prototypes by a master Designer is the strict evolutionist's adamant claim intelligence is absent from what appears to be beauty in an intelligently planned creation and life is nothing more than a happy accident.

WHAT IS LEFT TO DEBATE?

I should have clarified by now the stated objectives of both Richard Dawkins and myself from this debate. Mr. Dawkins wrote in *The God Delusion* "...if this book works as I intend, religious readers who open it will be atheists when they put it down. What presumptuous optimism!...of course, dyed-in-the-wool faith-heads are immune to argument..." [27]

Notwithstanding the curious thought of atheism being an optimistic point of view, Dawkins reveals his goal is to drive the religious person away from religion, using pop-science and flawed logic as his weapons of choice.

Dawkins does not wish for a debate on evolution, he wants a forum to bash religion. I hate to be the bearer of bad news to Richard, but reading his book failed to make me an atheist. He didn't even come close. I'm not sure what a "dyed-in-the-wool-faith-head" is, so I won't respond to that accusation. The skeptical foundation from which I developed my personal belief in God seemed to make me the perfect opportunity for Dawkins to win a convert, but railing against God does not make for a very compelling argument. Dawkins' goal to convert every religious reader to atheism was certainly a lofty goal.

I suppose my goal should be to convert Dawkins and every atheist fan of his to embrace my hybrid theory, but to quote Clint Eastwood, "A man's got to know his limitations." [28]

Evangelism is not my strong suit.

There's a saying which goes "You can lead a horse to water, but you can't make him drink." My opinions are offered for public consumption, and I invite all honest debate. Divine Evolution is something that has actually developed from the process of writing this book, so it's new. Please enlighten me wherever my logic or understanding of the facts goes astray. I enjoy thought provoking dialogue.

My strength is in using my mind to research and evaluate evidence, then to attempt a convincing counter argument to Dawkins' assertion that you are insane if you believe in God. Watching the Dawkins/Colbert interview and reading *The God Delusion* felt like the old chivalrous challenge had been issued, and I'd been slapped on each cheek with a glove. I knew what I already believed to be true about my God, but for a short period of time, I thought there might be some major discoveries of fossils to justify the bravado and confidence Dawkins exuded. There is not.

There is a plausible theory, Divine Evolution, which relies on scientific evidence but still acknowledges the critically important role the Creator played in designing and initiating creation. The debate left unsettled is no longer scientific, but logical and rational by nature. It is the debate of religion versus atheism, because to reject this hybrid theory requires denying any proof may exist beyond the current scope and ability of testing by scientific method. Because speciation and abiogenesis are also unproven via scientific method, Divine Evolution should not make even the pseudo-scientist squeamish.

5 EXISTENTIAL QUESTIONS

The sculpture known as *The Thinker* was created by Auguste Rodin. It portrays a man resting his elbow on his knee and his chin on the knuckles of his hand while he deeply contemplates some great mystery of life. Rodin originally titled his master work *The Poet* because the statue represented his vision of Dante as he paused outside the Gates of Hell to ponder a verse of what would become his *Inferno*, the first of three parts in his epic poem *The Divine Comedy*. Rodin's final version saw Dante literally perched on the Gates of Hell where he could meditate on the fate of those about to pass beneath him. The sculpture has become an iconic symbol that classically depicts man's struggle to apply intellectual thought to his predicament.

DANTE

The Divine Comedy followed Dante as he journeyed through the *Inferno* and the nine circles of hell. In the first circle of Hell, Limbo, Dante met the otherwise virtuous pagans such as Homer, Aristotle, Socrates, and his guide through the inner circles of hell, the Greek poet Virgil. Lust, the second circle of Hell, was where those consumed by sexual desires resided. Dante and Virgil saw Achilles, Cleopatra and Helen of Troy trapped in Lust.

Descending through levels of increased wickedness deeper into Hell, Dante and Virgil passed through the circles of Gluttony, Avarice or Greed, Wrath and Sloth, Heresy, Violence, Fraud, and Treason. The circle of Fraud was inhabited by those who knew and deliberately accepted evil, and Treason was where Dante met the betrayers Cain and Judas in the final circle of Hell the two poets passed through before encountering Satan at the very center of the earth.

After his journey through hell, in the second of the three sections of *The Divine Comedy,* Dante passed through Purgatory. Finally in the third part of his epic poem, Dante arrived in Paradise. Purgatory is said to represent the

Christian life on earth, and Paradise is of course Heaven, where Dante meets God.

The sentient human has long pondered the great questions about our existence, and how all living organisms came to have life. Why am I here? Is there a purpose for my life?

Frequently life is consumed by the mundane details of our day-to-day existence, and Sentient Man forgets to take time to reconsider the existential questions. Whenever new scientific or theological information comes to light, these questions should be asked again to see if new answers are possible.

It is easy to get caught in a monotonous routine with such a hectic schedule that one never takes time to meditate on these questions, much less to seriously work toward finding their personal answers. One may look into the stars at night and feel puny as a consequence. It is easy to watch a massive government grow out of control and declare we are powerless to stop it. As Edmund Burke said, "All that is necessary for the triumph of evil is that good men do nothing."

If we don't believe we make a difference, we lose hope. Those who have given up hope of gaining insight into their Creator by accepting the incomplete evolutionist theory as complete truth may feel that way. The rest of us stay busy looking for answers.

WHAT ARE THE QUESTIONS?

There may be others, but these are the most common existential questions we eventually ask ourselves. Whether the individual ever finds his answer is up to the seeker posing the question.

WHO AM I?

Everyone recognizes their own face in the mirror. Developing our sense of self is our first priority as we become conscious of the surrounding world. The development of our ego is influenced significantly by how the world reacts to us. Unfortunately society tends to overinflate certain egos. We treat professional athletes like they are rock stars. We treat rock stars like rock stars. Average people are well, average. The truth is rock stars and athletes are average people with one above average talent. Authoring a book does not make me any more special than you.

My theory of why movies like *The Bourne Identity* tend to be very popular is twofold: *The Bourne Identity* is a spy thriller starring a popular actor with exciting action scenes and stunts. However, several similar spy movies were made without enjoying the same success. The main hook for *The Bourne Identity* is the theme of lost memory and search for personal identity.

The plot of *The Bourne Identity* barely scratched the surface of the "Who am I?" question compared to the movie *Memento*. The main character of *Memento* lost both his identity and his short-term memory. He remembered his wife had been murdered and the injury that caused his forgetfulness, even the name of his wife's killer. But he couldn't remember which hotel room he'd rented without a numbered key as a constant reminder.

His unique solution for his perpetual memory loss was to tattoo clues on his own body as his search progressed for his wife's killer. It was one of the most remarkable films I've ever seen.

Our self-awareness begins in childhood as parents play the game "Peek-A-Boo" with their newborn infants. The game teaches the baby that Mom or Dad still exist even when they are not visible. It gives the child joy from the game and comfort in a subliminal sense.

One of my favorite characters in Douglas Adams' *The Hitchhiker's Guide to the Galaxy* was the Ravenous Bug Blatter Beast of Traal, described as one of the most vicious animals that live in the universe. The best defense against the creature is to wrap a towel around your head because the Bug Blatter Beast is "a mind-bogglingly stupid animal, it assumes if you can't see it, it can't see you." Apparently, if you fail to react to the sight of the creature, it would starve to death because it is "daft as a brush, but very, very ravenous." [29] The animal's weakness is due to a complete lack of sense of self.

I always thought the Bug Blatter was hysterically funny – until an incident occurred one evening while trying to put our youngest child to bed. His bedtime came before his older sister's, and he hated going to bed first. He loved to stay up and play in her room until her later bedtime.

One night I caught him trying to get into her room several times and as I watched, sure enough he stood in the doorway about to sneak across the hall yet again. When I informed my son I could see him, he slapped his little hands over his eyes and dashed for her room. Apparently my two year old son thought me as daft as the Bug Blatter, which did enormous damage to my self-esteem.

Interestingly, the Creator God is called Yahweh in the Old Testament because He gave that name when He called Moses at the burning bush to lead the people of Israel. When Moses asked God how to respond when the Israelites asked the name of the God who sent him, God said "I AM who I AM." Yahweh translated from Hebrew is "I AM." God has total and the ultimate sense of self.

Richard Dawkins also has a terrific sense of self, to the point of making sure the lettering of his name on the cover of *The God Delusion* was a half inch larger than the lettering used to print the word God, the unsubtle point being he is bigger than God.

In his interview with Stephen Colbert, Colbert greeted Dawkins with a joke, saying he was confused because he expected former NBA basketball star Darrell Dawkins.

Richard attempted a joke in reply, saying people often confused him for Steven Hawking, who is disabled. Dawkins said people expect to see someone who has difficulty speaking and travels by wheelchair. Dawkins is apparently trying to suggest he is as smart as Steven Hawking in an effort to claim intelligence similar to guilt by association.

It's kind of funny, because when I first heard his name I thought he said "Richard Dawson." I wondered what the star of *Hogan's Heroes* and *Family Feud* had to do with writing a book about God.

Of course, who am I to talk? When I used to introduce myself in a crowded room, someone would invariably ask "Did you say John Lennon?" and I would have to explain that no, I did not think I was a former Beatle, nor was I named after one. The names only sound similar.

We know the name we are given and our basic physical description, but does that tell us who we really are?

HOW DID I GET HERE?

At some point in our lives, we all must wrestle with the question whether we were created or evolved. Even the least curious mind must ponder the question occasionally. Despite the vociferous objections from Richard Dawkins, the truth is that parts of evolutionist theory require random chance to occur with invariably favorable results consistently. Dawkins is extremely sensitive to natural selection being lumped into evolutionist theory in that respect.

The Big Bang Theory is a conundrum of random chance and precisely tuned variables, which only works when there is an available catalyst to initiate it, and Dawkins knows it. This is why he speculates the so-called Rees' fine-tuned variables must have actually been constants.

The argument that the universe must be as it is because no alternate settings for Rees' variables exist is pure conjecture. Abiogenesis is pure conjecture. The dictionary describes it as discredited. Speciation is purely specious speculation. Natural selection is nothing more than the common sense observation of mutations due to inherited characteristics through genetics.

Creationist theory is simple. God created everything, and He did it in six days.

Divine Evolution looks at the evidence for the Big Bang and sees the only thing lacking is God as its catalyst; at abiogenesis and sees nothing without God's breath of life; at speciation and sees intelligent design; at natural selection and sees common sense. This hybrid theory marries the overly simplistic strict traditional creationist account to the evolutionist account and produces the only really viable theory from that union.

Do I have a purpose, or why am I here?

Pastor Rick Warren wrote *The Purpose Driven Life* to answer that question in the affirmative from the creationist perspective. Mr. Dawkins tells the story in *The God Delusion* of having lunch with a colleague who stated flatly, there is no meaning or point to our existence other than to take up space and eat a good lunch. The man is certainly entitled to his opinion, but I wonder if he also shares Dawkins' curious affinity for conscientious objectors (C.O.) to service in the military.

For people who consider their lives so worthless, it seems contradictory to place such a high value on living. The atheist C.O. wishes to preserve the life he considers so meaningless and is unwilling to serve and help protect his freedom to worship or not worship the God of his choosing. Without opposition, Sharia law would eventually be imposed worldwide by radical Islamic terrorists, leaving the "enlightened atheist" forced to make the unpleasant choice of converting to Islam or face imminent death.

Dawkins lauds himself for his enlightened perspective as an atheist and conscientious objector and will certainly object on the grounds as a C.O. he places a higher value on life and not killing others. It sounds noble, but then Dawkins doesn't disguise his admiration for the "ethics" of Peter Singer, which are radical eugenics in disguise.

The C.O. might even have the audacity to denigrate the service of those who do fight to protect his freedom and the same meaningless life the atheist feels isn't worth living. It is someone else's life the atheist typically deems no longer worth living, not his own.

If that sounds confusing, it is.

I happen to agree with Dr. Warren regarding his thought that we all have a purpose for being here. For a little while, I thought my purpose in life, besides raising my family, was to volunteer with our local Humane Society. My wife got me hooked on fostering animals that temporarily need shelter due to overflow conditions at the facility.

Over the past three years, we have fostered over fifty dogs and ten cats, successfully placed them into loving "forever" homes as each would find its perfect place in this world. A few dogs managed to avoid adoption and were grandfathered into our pack, which now numbers four additional permanent members of the family.

I have watched hundreds of hours of the amazing work of Cesar Millan on the National Geographic channel on television. Cesar, AKA *The Dog Whisperer*, rubbed off enough that I have become the resident dog rehabilitator for our Humane Society. That became my idea for purpose in life. Our shelter is a no-kill facility, which means we do not euthanize animals unless they are too sick, too badly injured, or too vicious to be placed.

Occasionally, good dogs begin to get stir crazy, if too much time goes by without adoption. A long-term shelter dog may begin to develop behavior problems. A dog may show what appears to be an aggressive personality but actually originates from fear or frustration. Dogs need a sense of belonging. They don't have a pack of their own at the shelter. Over time, these animals can be "institutionalized" by the shelter environment and become a problem.

Removing a troublesome dog from that environment to a more stable one (like my house) gives me the opportunity to work with the dog one on one and incorporate him or her into my pack. The most recent example is Balto, who happens to be lying at my feet at the moment. The bite wound on my wife's leg (she was his victim at the shelter), which still hasn't healed, came as she attempted to keep a frustrated Balto from going after another dog.

Most people would have decided to euthanize Balto after being bitten, but my wife recognized he was a good dog who had been changed from living at the shelter too long. She shuffled another foster dog out to the shelter so she could bring Balto here for me to rehabilitate. This helped both dogs, as the one who left was adopted quickly with his increased exposure to shelter visitors.

Balto has transformed into a happy and stable dog that enjoys his walks with our pack. There is no doubt in his mind or mine that I am pack leader. One quick command is all that is required for Balto to roll on his back so his belly can be scratched. It is an extremely rewarding experience to watch him transform and then find a wonderful dog like Balto a great home. If I could keep him, I would, but I say that about every dog that passes through here, but if I did that, we would have 54 dogs today.

I come to love them all, and I miss each of them when they find forever homes. My wife and I have come to realize we can only keep the broken ones who bond too strongly to us and would not do well if they left. We learned that from Rusty, our collie-German Shepherd mix who was adopted thrice but returned every time for repeated behavior problems that he never shows at my house for some strange reason. He made his point, and he will not be adopted by anyone else. Rusty adopted us.

I was perfectly happy with the thought that working with these animals was my true purpose in life until I began to write this book. Now I think I have more than one purpose. I think this book is important not because I wrote it, but because God wanted me to write it. I don't ask for credit for this work because God is the One who deserves it. I've just sort of been along for the ride.

IS THERE SOMETHING GREATER THAN MYSELF?

Are we hard wired to seek God? Every culture seems to have some sense of at least one god; there has never been an atheist culture to the best of my knowledge. Atheism begins to spread in cultures that adopt false gods after they are proven false. Richard Dawkins might want to believe that God is a psychological creation attributable to memes, but the thought has to originate in order to be able to replicate. Where did the idea for God originate, if He did not really visit Moses on Mount Horeb in the form of a burning bush?

I believe with strong conviction that God does exist, while Mr. Dawkins is of the polar opposite opinion. My belief is based on three things: my personal experiences, observations and knowledge acquired through the course of my lifetime, and my recognition that there are things that can never be known, defined, or comprehended. There are things in this world which may never be understood. These things cannot be measured in cubic centimeters or kept in a test tube. That does not diminish or mitigate the truth of the unknown and unknowable.

Evidence of God's brilliance, His creative genius, and the awe-inspiring works of His hands are ubiquitous.

My previous house had a small pond in front, and from time to time I enjoyed sitting on the front steps and admiring God's creation. Blue herons waded in the shallows to fish for blue gill. Hawks circled overhead in search for a wayward squirrel or sparrow. Pine, oak, magnolia, mimosa, maple, and river birch trees all colored the skyline. Rabbit and deer sometimes spooked by my dog pack raced for the safety of the woods on either side of our front yard.

Life was diverse, abundant, wild and domestic. Flora grew plentifully where man and woman planted, and ferociously where we neglected to clear and claim. As I sat and marveled at God's creation, I saw obvious evidence of intelligent design. Details that required planning and forethought graced the landscape everywhere I looked. Dawkins is fond of stating the overwhelming appearance of design is misleading, but the truth is the appearance is overwhelming because it is reality.

The evidence for design is present both on the individual level when observing a single example of any species as well as on the macro level when observing how life forms work in conjunction and cooperation with each other. The fish in the pond eat insects and smaller fish. Birds eat fish and snakes. The snakes hunt small rodents. Hawks hunt them both. A small ecosystem or miniature food chain is all right there in one place.

You can dig a pond and it may become stocked with fish without ever adding any. As birds stop to drink, fish eggs from other places that rode along may hatch in your pond, and voila, you have fish. Once fish get into the pond, creatures that like to eat fish will be attracted to it and a small ecosystem is born. Life is never spontaneously generated by accident, and life attracts other life. If poison is introduced into the pond and all the fish are killed, the life that was attracted will be repelled. If you want to call it a law of nature, you can, but its common sense more than anything else.

My personal version of creationist theory (Divine Evolution) holds that the earth is old. God created the universe, the Earth, and life on Earth over an unknown period of time when measured by Earth days. The length of time it took for creation is unknown because we don't know how long a day is to God.

The Big Bang occurred when God spoke and said, "Let there be light," and matter was created at that point. Matter did not precede the Big Bang in the dimensions of time and space we know and understand.

Abiogenesis occurred as God breathed Life into inanimate matter and speciation occurred over eons of time as God tested and experimented with His creations. God, the artist, periodically cleared his palette of obsolete prototypes and executed production of His modified design. Natural selection is only common sense to believe because there had to be a mechanism provided by God to create modern life post Noah's Ark, not to mention the five mass extinctions known as the Cretaceous-Tertiary extinction, the End Triassic extinction, the Permian-Triassic extinction, the Late Devonian extinction, and the Ordovician-Silurian extinction.

These five instances all wiped out significant portions of both terrestrial and marine forms of existing life. The Permian-Triassic extinction, which occurred roughly 250 million years ago, was the worst mass elimination of life on the planet in history with an estimated 95% of all species on earth wiped out. Theories believe the extinction was due to a comet or asteroid strike on Earth or possibly caused by flood volcanism.

The latter theory would involve a temporary change in the Earth's atmosphere that led to the loss of oxygen in the seas. Either way, the "slow, gradual process" required by the theory of speciation to work is disproved by the real fossil record. There are multiple instances of the eradication of large portions of all existing life on earth. All the dinosaurs were wiped out in one fell swoop during the Cretaceous-Tertiary extinction that occurred about 65 million years ago. [30]

WHAT IS DEATH? WHAT HAPPENS AFTER DEATH?

The biblical account of the life of Jesus requires the creationist to accept the concept of life after death. Jesus said, "I am the Resurrection and the Life," and the gospel accounts post crucifixion agree he rose from the dead and ascended into heaven. Jesus also raised Lazarus from the dead to demonstrate fully his power over physical death. But what about for the rest of us?

There is supporting evidence of belief in the four eyewitness accounts in the New Testament known as the Gospels of Matthew, Mark, Luke and John. Three of the four accounts were written by participants in the events as they occurred, and the fourth was credited to a physician who joined the fledgling religion and based his account on testimony obtained from numerous eyewitnesses. But is evidence of Christ's life limited to the Bible?

General details of Jesus' life and crucifixion were recorded by the Jewish historian Josephus and the Roman historian Tacitus. There is no reason to question the existence of the historical Jesus. What about the mystical Jesus as Son of God? Because of my personal experience, I also believe Jesus is the Christ.

Jesus did say, "I am the way, the truth, and the life. No man comes to the Father but by me." [31] But if we interpret that to mean only Christians go to heaven, do we risk being judgmental? Jesus also said, "Judge not, and you will not be judged." [32]

Does John 14:6 mean we must all be Christian to go to heaven, or does it mean Jesus will be our advocate and judge? Using the advice Jesus offered Matthew 7:1, I am no more qualified to judge the religion of another anymore than I am to judge an individual believer.

Who am I to judge that the devout Hindu, Jew or Muslim is unworthy of heaven? I will always be happy to peacefully discuss my beliefs about God with anyone of any faith, provided we each agree to politely listen to each other's point of view with respect.

These are the most relevant existential questions I have asked myself, or at least every one I could think of. I have spent a fair part of my life, so far, contemplating these questions and the different possible answers. The convictions I currently hold are strong, but they are not definitive and conclusive, nor do I believe my answers can ever be during my lifetime. The only way you could say for sure there was life after death is to die, making it exceedingly difficult to prove your point.

Richard Dawkins alluded to the existential questions, but he didn't want to dwell on them too much because it detracted from his full scale assault on God, and he probably recognized he would lose some of his audience if he emphasized the hopelessness and emptiness of atheism.

Dawkins uses two methods to deal with the existential questions. First, he quotes heavily and relies upon the philosophy of Bertrand Russell to make his argument against God, which is what the debate has now degenerated to become. Second, Dawkins attempts an interesting trick; he blames the creationist for what he does routinely, apparently borrowing from the Vince Lombardi school of philosophy, which taught as rule number One "The best defense is a good offense."

Lombardi also said "Show me a good loser, and I'll show you a loser." Vince might have been a great football coach, but he certainly was not the ideal role model for good sportsmanship. It also proves not every maxim he issued is applicable to life outside of football, either.

In *The God Delusion*, Dawkins expressed great admiration for Bertrand Russell's logic and reasoning, citing as a favorite example his celestial teapot parable. Dawkins quoted that Russell wrote "If I were to suggest that between the Earth and Mars there is a china teapot revolving around the sun in celestial orbit, nobody would be able to disprove my assertion provided I were careful to add the teapot is too small to be revealed even by our most powerful telescopes. But if I were to go on to say that, since my assertion cannot be disproved, it is intolerable presumption on the part of human reason to doubt it. I should rightly be talking nonsense. If, however, the existence of the teapot were affirmed by ancient books, taught as the sacred truth every Sunday, and instilled in the minds of children at school, hesitation to believe in its existence would become a mark of eccentricity and entitle the doubter of the attentions of the psychiatrist in an enlightened age or of the Inquisitor in an earlier time."[33]

Unfortunately for Dawkins, by quoting that passage from Russell, he opened my eyes to the obvious fact evolutionists have not one but two celestial teapots–abiogenesis *and* speciation. Neither can be proved but both are taught as the sacred truth of evolutionary theory from their sacred tomes of science textbooks in science class every Monday through Friday, and both abiogenesis and speciation are utter nonsense.

Abiogenesis cannot be disproved because it is claimed the conditions of the original Earth atmosphere cannot be duplicated today. Speciation cannot be disproved because of the time factor required for species to split or merge. But the creationist can (and should) rightfully argue that abiogenesis can't ever be proved because the primordial soup itself has no rational explanation for being outside of God creating it. The only thing that really matters is this: where did matter originate? This stuff we call flesh on our bodies, sand on the beach, a daffodil–it's all different kinds of matter, but it's all made of matter. The bigger question than who made them is who made the raw materials from which each is constructed?

And the word "constructed" was of course deliberate because the word implies intent and design, which is obvious to an unfortunately shrinking percentage of the world's population. It's trendy and politically correct to be atheist, agnostic, or even downright hostile and nasty toward religious people, especially those categorized as Evangelical Christians.

Evolutionists declare it is intolerable presumption to doubt either theory, because to question them is to insinuate that God exists. Evolution is taught as sacred truth every day school is in session. Because of Dawkins' affinity for Russell's philosophical point of view, which seemed to influence his thinking so powerfully, I checked out a collection of essays written by Russell titled *Why Am I Not a Christian* to learn more of his philosophy for myself.

WHY I AM A CHRISTIAN (WITH NO APOLOGIES TO BERTRAND RUSSELL)

Bertrand Russell wrote the essays included in *Why I am not a Christian* to provide what he hoped would give free thinkers a solid rationale to reject faith in God. Reading his now mostly antiquated arguments made it easy to pick apart the logical flaws in his philosophical point of view. It also makes me curious why Dawkins finds them so important because so much of his logic was flawed given modern evidence. Russell based several key arguments on scientific theories no one believes or takes seriously anymore.

Russell and Dawkins both spent a fair amount of time and effort attempting to refute the "ontological argument" about God, which states if humans can conceive of God, then He must exist. Russell criticizes (and

Dawkins agrees) the ontological argument as circular in nature. Russell also claims the argument commits a bare assertion fallacy, meaning the Bible presents an argument that is assumed to be true because religion says it is true.

I say the ontological argument itself is ridiculous. Humans didn't just dream up the idea of a Creator to control the masses. The Sentient human was either informed of God's existence by God as the Bible described, or Adam, Moses, Abraham, Joshua, Peter, Paul and even Jesus all separately suffered from the same delusion I do about the existence of God. Even if Richard Dawkins wants to insist some of these men never lived, he would never suggest all are fictional characters.

I cannot speak for anyone but myself, but I believe God to be true because I really wanted to know and so I looked for myself. I was skeptical. Dawkins and Russell are not skeptical but cynical. Whether it is due to personal prejudice born of experience or, whatever reason, their bias against God overwhelms any possibility of honestly considering Him for a moment.

Russell may have something of a point about the bare assertion fallacy, but is not the converse argument also true? Doesn't the evolutionist argument assume the Bible is false because they believe so strongly that evolution is true? It is critical to note because the evolutionist does not believe in the young earth or six days of creation, all of creationist theory is rejected. On what some might consider a technicality, an entire system of belief is rejected only to be replaced by an even more unbelievable and fantastic explanation for life. The latest theory about the origin of life on Earth (without God) is that it was brought to Earth by an asteroid or comet.

But just as the evolutionist is quick to point out if God made something, then someone greater than God had to create God, if life came to earth on a comet, what created the life on the comet? Where does life come from? Shifting the source or origin off-site is a classic "C-Y-A" maneuver; it opens up new avenues to argue that haven't been disproved yet to replace the old ones that are now trashed and recognized as obsolete. Creationism remains discredited in their minds due to a misunderstood or misinterpreted technicality. The die-hard evolutionist who refuses God a role in creation must continually invent new excuses for life because there isn't a rational explanation for how it came to be apart from God as the instigator.

ALPHA AND OMEGA

If you want to talk about circular logic and a bare assertion fallacy, look no farther than the atheist's argument against what is called the "First-cause" argument by Bertrand Russell in his titular essay *Why I Am Not a Christian.*

Russell quotes John Stuart Mill as saying, "the question 'Who made me?' cannot be answered, because it immediately suggests the further question 'Who made God?' If everything must have a cause, then God must have a cause." [34] The irony of arguing the ontological argument for God is circular using the argument that a Creator requires a Creator. We infer a Creator is responsible for the natural world and what we observe. Because we cannot observe the realm of the supernatural, we cannot infer any natural laws apply to a supernatural God.

Russell's effort to crush the "First cause argument" actually accomplishes the opposite. He says "There is no reason why the world could not have come into being without a cause, nor, on the other hand, is there any reason why it should not have always existed. There is no reason to suppose the world had a beginning at all. The idea that things must have a beginning is really due to a poverty of the imagination. Therefore, perhaps, I need not waste any more time upon the argument about the first cause."

Russell fails to grasp the concept that God is ageless, timeless, and endless, and therefore origin-less as well. There is no rule that declares God must have a beginning if everything else did except in the constrained minds of some less erudite philosophers.

But that is the only argument the evolutionist has left against God at this point in respect to the "first cause" existential question at hand.

Yet we know there was a beginning to the Universe, and an origin to life on Earth. Russell based his entire argument against the first cause on the now debunked static state theory, which says the universe was always here in the same basic form it has today. Einstein might have agreed with Russell back in his day, but no serious scientist (or pseudo-scientist) today believes the Big Bang didn't happen.

The Big Bang Theory is now the only current scientific theory taught that describes the origin of the universe. If the static state universe theory is ever taught in science class, it's only for a historical perspective and not as an actively valid theory. It is no longer taken seriously.

Russell's argument against the First Cause was not first-rate. It was lame and pitiful because modern science refutes it. The centerpiece of his argument is no longer considered a viable theory. Therefore, I need not waste any more time with his argument.

Creationist theory is attacked as a single monolithic argument, so any perceived flaw in the theory is seen as discrediting the entire theory. By contrast, evolutionist theory can be separated so individual parts may be called true and then the whole theory becomes accepted by extrapolation. It is easily defended because its attackers call it evolution rather than naming the specific theory in question, allowing the evolutionist his opportunity to change the topic.

Russell reaches too far in his effort to disprove the natural law argument for God when he says, "[W]hen you get down to any knowledge of what atoms actually do, you will find they are much less subject to law than people thought, and the laws at which you arrive are statistical averages of just the sort that would emerge from chance. There is, as we all know, a law that if you throw dice you will get double sixes only about once every thirty-six times, and we do not regard that as evidence that the fall of the dice is regulated by design; on the contrary, if the double sixes came every time we should think that there was design."[35]

His point is downright silly. Simply the fact two identical cubes exist with the exact same markings on every side smacks of design. If double sixes were to come up every time for thirty-six straight rolls, I would suspect cheating and believe that the dice are loaded. My fundamental understanding of the laws of probability tells me the chance of that many consecutive tosses rolling double sixes are so low it may as well be considered impossible. In fact, it wouldn't take thirty-six rolls to reach the conclusion the game was fixed. I would suspect something was wrong if consecutive turns throwing double sixes reached double digits.

Loaded dice are a product of dishonest but intelligent design. Try telling a physicist his natural laws are only statistical averages that emerge from chance.

LONG WAY DOWN

As one scientist has noted, if you doubt the laws of gravity are real, just make sure you don't first test your hypothesis from the top of a tall building. It's better to prove your theory gradually rather than taking a giant leap of faith.

The fall may not kill you, but that sudden stop at the end often does.

That reminds me of the very first Monty Python skit I ever watched. One night while still in college, I was eating a late dinner after work. I flipped around the channels on television looking for something interesting to watch.

I found a program where two guys were just standing in a field doing nothing for several seconds. Curiosity got the better of me, so I took a bite and watched to see why these two men stood in a field saying nothing to each other while I chewed my food. Both men were looking off camera and upward, saying nothing. Within a few seconds, they looked down rapidly as if they had seen something fall.

Then one man asked the other "Pardon me, but do you have any idea why those sheep are up in those trees?"

I was hooked.

The other man launched into a lengthy tale of a sheep named Harold, who had led a mutiny of sorts by the other sheep. Harold decided sheep had a miserable routine of being sheared and herded, and ran the risk of being served for dinner any given time. Harold and the other sheep decided they quite liked the freedom and lifestyle enjoyed by birds. This led to the sheep climbing into the tress in order to learn how to fly. The punch line was then delivered: "But the interesting thing is that sheep do not so much fly as plummet."

Russell concludes his argument against natural law by saying "The arguments that are used for the existence of God change their character as time goes on. They were at first hard intellectual arguments embodying certain quite definite fallacies. As we come to modern times, they become less respectable intellectually and more and more affected by a kind of moralizing vagueness." [36]

This argument could be used equally effectively against him by merely substituting the word "evolution" for God, because as we have already seen, the term evolution is a deliberately vague term to cover four separate processes; the origin of the universe, the origin of life, speciation, and natural selection. All four theories must have plausible arguments before one might legitimately eliminate God from the equation.

The logical circles made by these arguments remind me of a comic scene from Douglas Adams' book *The Hitchhiker's Guide to the Galaxy*. A fictional space creature called a Babel fish was used by man to successfully argue God from existence. The Babel fish is described as a creature a person could put into their ear, and the fish would translate alien languages into the native language of the listener. The argument that got rid of God went like this:

"I refuse to prove that I exist," says God, "for proof denies faith, and without faith I am nothing."

"But," says man, the Babel fish is a dead giveaway, isn't it? It could not possibly have evolved by chance. It proves you exist, and so therefore, by your own arguments, you don't. QED."

"Oh dear," says God, "I hadn't thought of that" and promptly vanishes in a puff of logic.

"Oh, that was easy," says man, and for an encore goes on to prove that black is white and gets himself killed at the next pedestrian crossing. [37]

Adams' humor pokes gentle fun at the mental gymnastics caused by logical thought conceived by the atheist mind. By excluding God from the creation debate at all costs, we eliminate our most obvious explanation for life.

Russell argues against design with a particularly audacious comment: "When you come to look into this argument from design, it is a most astonishing thing that people can believe that this world, with all the things that are in it, with all its defects, should be the best that omnipotence, omniscience and millions of years to perfect your world, you could produce nothing better than the Ku Klux Klan or the Fascists?" [38]

BOTTOM OF THE FOOD CHAIN

The best answer to this argument first refutes his premise the world lacks 'proof' of design. Then what explains the food chain? We aren't just talking about design within an individual creature. We can see obvious purpose for existence in the existence of multiple living organisms that provide a source of food for other animals. Why would an animal evolve only to become the perfect meal? Are we to believe that plankton evolved only so they might be eaten by whales? The implication of the evolutionist argument is some plankton had the sense to become whales, and the rest became cuisine.

So where does this leave us?

The search for answers to the existential questions is a journey, not a destination. We can stop looking when we take our final breath. From that point forward, we should be prepared to begin a new journey.

The best way to search for these answers is to learn from our personal experiences.

6 PERSONAL EXPERIENCE

Let me tell you about the night I met God. If you are familiar with the expression "be careful what you ask for because you just might get your wish," this story will make perfect sense to you.

I was lying in bed one night as my wife slept next to me. I couldn't sleep. As I tossed and turned, I wondered, "What if it is all a hoax? Could religion be the greatest practical joke ever played in the history of man?" I asked myself if God could be a figment of our collective imagination.

Of course, by God I meant Yahweh the Creator God, who was even harder to believe in than Jesus. There were eyewitness accounts of Jesus' life in the Bible, but I was unaware if any external documentation existed of his life. I wondered if the Old Testament stories about God were only fictitious stories part the plot of this book called the Bible. I was taught the Bible contained truth as a child, but then I was taught it was fiction in college. Which was true? At least the side advocating evolution had a little evidence on their side as opposed to none to support beliefs of religious people, or so I thought....

So, there I was, tossing and turning. My mind raced to consider one possibility after another. I remembered a scene from a movie called *And Justice for All* that starred Al Pacino. In the climatic courtroom scene, the defense attorney played by Pacino kept asking the jury why the victim had accused his client of assault, battery and rape.

His client was a distinguished judge who was well respected in the community, so Pacino kept repeating, "Why would she lie?" over and over as he mentioned one motive after another before he dismissed each as a possible reason for her accusations. Finally he thundered, "She isn't lying. The S.O.B. is guilty!"

Of course, in the real world, such a scene would result in the immediate declaration of a mistrial for the defendant and probably the initiation of disbarment proceeding against the attorney, but it made for great cinema.

That same question seemed relevant to me as my mind applied it to the Gospel writers–why would they lie? What possible motivation could Jesus' followers have to create a new religion to follow his teachings after he died on the cross? It surely wasn't profitable financially in the early years; in fact, the reward of many was persecution and death for themselves as well. Of the disciples, only John died a natural death, and he survived being boiled in oil. Why would these men lie? What did they stand to gain?

IS IT REAL?

As a child, I had been conditioned to believe in Santa Claus and the Easter Bunny. Santa Claus supposedly delivered Christmas presents to every good boy and girl on Christmas Eve and the Easter Bunny delivered chocolates and Easter eggs for Easter Sunday morning. As an adult, I knew Santa Claus and the Easter Bunny were fictitious characters, because I also believed that it was a role I was supposed to play for my own children.

My wife is an independent thinker, which helps explain why she agreed to be my partner for life. When our children were very young, she quietly refused to engage in the traditional anticipation of Santa Claus. Our children understood the presents we gave came from us, which effectively deprived us the opportunity to elicit good behavior as bribes for a better haul at Christmas from Santa. The children still got plenty of Santa frenzy from the grandparents who insisted Santa visited their house instead. But their mother adamantly refused to play the role of Santa Claus.

One night she explained her reluctance to me after she tucked the children in bed. "If I tell them now that Santa Claus is real, how can I expect them to believe me when I talk to them about God? Why should they believe me about Jesus, if I lie to them now about Santa and the Easter Bunny?"

Needless to say, these Christian folklore stalwarts of American childhood were understated in our home, better left to grandma's house. Having been raised in the more traditional Christian home, I still hold the bitter memory of the day I learned Santa was imaginary and have to admit it didn't help my belief in God.

After I learned the legend of Santa Claus developed from the true story of St. Nicholas, the thought occurred to me Nicholas would be mortified were he alive today. The mythical Kris Kringle or Santa Claus was born in American culture due to a language barrier. European immigrants celebrated the gift giving tradition started by Saint Nicholas as part of Christmas festivities.

Their new neighbors appreciated and wanted to copy the tradition. Nicholas had evangelized to Europe, and as part of his ministry, to create anticipation and excitement for the birth of Jesus, he crafted small toys he could give away during the Christmas celebration.

When American immigrants observed the tradition and began to copy it, the question was asked about the origin of the gift giving, and the heavily accented immigrant's response of "St. Nicholas," when pronounced "Saint Nick-Loss," sounded like "Santa Claus."

That fateful night I pondered whether the disciples perpetuated a fantastic ruse after Jesus was crucified. History offered me knowledge of other rebellions and cults, which practically disappeared overnight after their leader was killed. But Christianity is unique among religions developed as movements following a charismatic leader.

It gained steam *after* Jesus was executed. The dominant Roman culture methodically tried to eliminate the sect in the first few hundred years after Christ's time on earth, yet the opposite effect occurred. The early Christians had many martyrs, but for every Christian put to death, it seemed four new converts took his place. And what could possibly explain the transformation described by the story of Saul of Tarsus?

After Jesus was crucified, the early church had two great leaders emerge. Peter was one of the disciples and evangelized the Jews. Jesus told Peter that his faith was "the rock" on which the church would be built.

By contrast, Saul of Tarsus was the chief persecutor of the early Christians and a religious leader of the Jewish community. Prior to a fateful journey between Jerusalem to Damascus, Saul was best known for presiding over the stoning death of St. Stephen, a martyr from the earliest Christians.

On the road to Damascus, a Light temporarily blinded Saul. A voice, only Saul could hear, directed him to a street named Straight and the house of Judas (not the disciple), where Ananias was also led to restore his sight. The experience had such a profound effect on Saul that he changed his name to Paul and became the second great evangelist leader of the early Christian church. Paul wrote most of the books in the New Testament in the form of letters to various early churches.

While these confused thoughts and remembered facts swirled through my head, I also reflected on a few Bible verses I was required to learn as an adolescent in confirmation class. Expecting a teenager to remember anything not related to sports, pop culture, or the opposite sex is a leap of faith in itself, but somehow these verses came back to me in the darkness of the night.

THE DOOR

One verse was Matthew 7:7, which says "Ask and you will receive, seek and you will find, **knock and the door will be opened for you**." The other verse connected to that thought in my mind was Revelation 3:20, which says "Behold, **I stand at the door and knock**. If any man hears my voice and opens the door, I will come in and be with him, and he with me." The dots started to connect.

At that moment, I really wanted to know the truth. I prayed, very simply and earnestly, saying, "I'm open, please come in." I repeated it several times and sincerely meant it with every fiber of my being.

The thing is, when I look back on that night, I realize my mind was conditioned not to expect any response. I had prayed before and those prayers weren't answered, so I didn't really expect this one to be answered, either.

Of course, at the time I hadn't realized those other prayers were self-serving and useless. The most sincere of my previous prayers centered on getting me out of trouble or having something materialistic I wanted to come to fruition. My mind was conditioned to pray and expect nothing to happen.

So when the bedroom filled with a bright light I was stunned. I knew a God powerful enough to create the Earth should not stoop to visit my bedroom because I was unworthy.

I knew exactly what a selfish, dishonest, and petty person I could be; not the public persona most people seemed to like, but the real guy inside who was not so nice as I pretended most of the time. But where do you hide from God when He comes to see you? I closed my eyes and lay quietly playing possum in the hope God would ignore me.

But the Light didn't leave me alone; it washed over and engulfed me. I felt indescribable peace and tranquility. When I finally reopened my eyes, the Light had faded and darkness returned but I was not afraid anymore. Even with my eyes closed, the Light had changed me. It left me with a feeling of peace and tranquility I have found impossible to adequately describe. I can't say I've felt that peace as completely since that night, but enough residual effect has remained with me that my life changed forever from that night forward.

I'm not saying I became a completely different person who would have made Mother Theresa proud, but there is a difference I can tell. Before that night, I used to hate people. Certain people would do me harm or cause me pain or simply aggravate me, and I would just hate them.

Now, I only hate the actions of people and not the people themselves. I tend to feel pity for those perpetrators of evil deeds, because they do not realize they also hurt themselves when they inflict pain on others. We will all suffer consequences for our actions; it is only a matter of *when*, not *if* we face judgment.

Had I prayed that night and nothing happened, I would almost certainly be an atheist today. Perhaps that is why I was granted the experience that night. I suspect it was rather due to the intensity and sincerity of my prayer that particular night, but only God can say for sure. The complete peace of the Light will remain with me as long as I live and opened my mind to endless possibilities. The surreal and supernatural quality of that night changed my view of life permanently.

I can't prove through testimony in court that these events happened that night, but I could testify with a clear conscience I believe the events I just described really happened.

NATURAL AND SUPERNATURAL

Mr. Dawkins has made it very clear that he does not believe there is such a thing as supernatural gods or a supernatural God. But God can only be supernatural. The biblical definition of God requires superhuman traits to be called the Alpha and the Omega, the I AM, the beginning and the end of all we know right down to the particles of dust that blow in the wind. A physical presence implies the existence of a being constructed from matter. Since God the Creator made matter, it is only logical to assume He is not made of matter. God must surpass all laws of physics and science that apply to us.

The circular First Cause argument popular with the evolutionist says that if God created matter, the question must be asked "who created God?"

The assumption is that if everything else has an origin, God must also by extrapolation. However, we just established that God must exceed all physical boundaries our human minds perceive, otherwise creation could not have occurred. The argument God must have an origin is a specious one at best.

The meaning of the word supernatural is "above and beyond what is natural," isn't it? The dictionary also describes the supernatural as the "direct influence or action of a deity on earthly affairs." Supernatural creatures abound in the lore of every culture on earth. Is all of humanity delusional or might there be some factual basis or true origin to supernatural tales of vampires and ghosts?

For example, the legend of Dracula originated from the true story of Count Vladimir Dracul of Transylvania, more popularly known as "Vlad the Impaler" due to the grisly method of execution he preferred to inflict on enemy invaders of his land. Bram Stoker merely heard the true stories and crafted a sinister fable. The point is, I never believed in supernatural creatures such as vampires or ghosts because those were mythical creatures in Hollywood movies. Dracula was never a real creature, purely fictitious in my mind until I learned the true story of Vlad the Impaler.

Savannah, Georgia is the oldest city in the state, founded in 1733. Savannah has a reputation as the most haunted city in America, according to the American Institute of Parapsychology. Savannah is the place I called home during my adolescent years.

I didn't believe in ghosts until I became friends with a guy named Jim Nettles. Jim and I met while growing up in Savannah and found we shared common interests in sports, chess, and chasing girls.

Jim lived around the corner from a bar called Stonehenge, a popular disco at the time, which attracted plenty of single young women. I would stop by Jim's house after work, and we would head out for the clubs together.

Jim lived in what we native Savannah natives call the midtown area in a big house on a corner of Victory Drive. The house was built in the early 1900s. However, the house was retrofitted with a few modern conveniences like an elevator that was installed after Jim's father had a heart attack to make it easier for him to get up and down the stairs. Motion sensors were included as part of the security system protecting the house.

One night, as we were about to leave for the evening, Jim and I stopped in the kitchen briefly while I made each of us a shooter (a shot of alcohol with flavoring). I had just gotten off work for the night and was cold sober, a condition I sought to remedy. But let me be absolutely clear; not a drop of alcohol had passed my lips before the following event occurred.

Jim reached over and turned off the overhead light in the kitchen then moved several feet clearly away from the switch but remained in my field of vision. Only a dim light over the sink illuminated the kitchen. The overhead light operated on a three-way switch, but both switches were in my view. Any movement to touch either switch could not possibly escape my notice.

The overhead light came back on. It took a second to register the significance of what had occurred. I knew what I'd just seen happen, but my mind refused to believe it. I had visited the house often enough before that night to be familiar with how the light switches operated, and believed what I'd just seen was impossible.

The house where I lived had single bar switches. If you swatted at one it was conceivable the bar could go halfway and then "flip" back on its own. Jim's lights had old-fashioned two button switches. When one button was pressed, the other button would toggle. When the pushed button clicked, the lights would be off or on until the other button was pushed.

Jim and I were the only two people in the house at that time. Jim was the only person who could have conceivably played a practical joke on me. That's exactly what I thought he'd done. Jim is very intelligent, and I knew from previous observations of the switches one could not accidentally toggle off and on.

The logical conclusion was that Jim tampered with the switch so he could trick me. The only problem was I hadn't seen him do anything after he turned off the light.

As I walked nonchalantly toward the switch Jim had used, I casually said, "That was a neat trick, Jim. How did you pull that off?"

Jim's answer stopped me in my tracks. He just laughed and said, "Dude, my house is haunted."

At first, I didn't believe him. I went to the switch and checked it out, but there was no sign of tampering. It spooked me a little. Until that moment in my life, ghosts were either a cartoon character named Casper or the product of a delusional person's imagination. At least Casper was friendly.

I was experiencing the unthinkable, and my good friend acted as if it was perfectly natural and ordinary. Looking back on that night, I'm quite sure I remained somewhat skeptical and didn't really believe in ghosts after that first incident, but my skepticism had been given a healthy shake and the foundation had been laid for me to come to believe in them.

Over the course of several more years, I personally witnessed and experienced, on many different occasions, a variety of inexplicable phenomena at Jim's house. For example, one night Jim and I were playing a quiet game of chess in an upstairs sitting room. The rocking chair positioned off to the side began to rock as if a spectator had taken interest in our game. They rocked as they watched. I was distracted watching the chair rock at a steady pace, and apparently my lost interest in the game bothered our supernatural spectator. The rocker stopped moving suddenly, exactly as if someone had put their hands on the arms of the chair to stand up.

The sound of footsteps preceded what I sensed as the brush of movement so close behind me, I could feel the air disturbed around me. The hair stood up on the back of my neck.

Yet before Jim and I both left Savannah for college, I became accustomed to such occurrences. The more frequently I observed an unusual event, the more accepting I became of Jim's explanation.

The motion sensors would go off when no one else was home to trigger the alarms; the elevators rode up and down by themselves. Footsteps echoed down a hall indicating someone's approach when no one was there.

Too many weird things happened at the house that defied explanation, but they happened just the same. I wasn't the only one who couldn't say what caused all the strange things that happened.

Paranormal researchers studied Jim's house. They declared as many as five spirits lived there with Jim and his parents. A local author included a chapter about the house in a book of famous Savannah ghost stories.

But I must confess I never felt any animosity or hostility when I felt these "ghosts" or had interactions with them. I never saw anything more dramatic than the experiences that I've described.

From one story the family told, the ghosts were only friendly up to a point. They owned the house on the adjacent corner as well and rented it to tenants. The property was rented fully furnished to a young couple with several small children. The children were said to have been running and playing around the house wreaking havoc as small children are prone to do, when an invisible hand started at one end of the fireplace mantel and swept to the other end, sending books and bookends flying. The act appeared to be the result of a ghostly temper tantrum. At least, the new tenants interpreted it as such, and they moved out immediately.

That Jim's house was allegedly haunted has been documented in the book *Savannah Spectres*. The author confirms the Nettles hired at least three different electrical contractors to evaluate the house and determine the cause of the apparent motion sensor and elevator malfunctions. All three contractors failed to find anything wrong with the electrical wiring.

As humans, we tend to believe only in those things we experience with our five senses, and discard as fiction or fairy tale the phenomena we have not. Yet as the evangelist Billy Graham pointed out, "I've never seen the wind; I've only seen the *effect* of the wind."

I have no problem believing in wind because I've experienced it, as have we all. I experienced something not everyone else has, but I still have no problems believing the experience was real.

BONAVENTURE CEMETERY

Jim's house opened my eyes to the supernatural, but it is not the only reason I believe in ghosts. Bonaventure Cemetery in Savannah was famous long before the book *Midnight in the Garden of Good and Evil* by John Berendt and the movie of the same name directed by Clint Eastwood. I had an interesting experience there in Bonaventure one night. I'm not quite as clear what happened there as I am about the experiences at Jim's.

Tad, my friend from school had a crazy idea when he picked me up after work one night. We were bored and trying to decide what to do next, so Tad suggested we investigate the local legend that Bonaventure cemetery was haunted.

Tad drove my car toward the back section of Bonaventure called Greenwich. Greenwich sits on a bluff on the Bull River. It is very spooky at night when fog rolls off the river and swirls around the Spanish moss hanging from the massive oak trees that line the road and the bluff.

We passed through the wrought iron gates that mark the entrance to Greenwich, which rumor said was the entrance to the grounds of the original plantation before the house burned down and the property was converted to use as a cemetery.

Suddenly Tad acted very excited. He said "There's something moving right outside the car! It's on your side. Do you see it?" I looked out my window but saw only darkness.

My initial thought was he was pulling a prank. The whole idea smacked of a ruse. But when I heard the fear in his voice, I knew my friend wasn't that good of an actor. Tad really was scared.

But of what? I still didn't think anything was outside the car. Admittedly, the atmosphere of the cemetery created a scary aura, but looks won't hurt you.

We only drove approximately one hundred yards past the gates. Tad whipped the car into a u-turn as soon as the concrete island dividing the road entering this section of the cemetery came to an end. We headed back to the gates.

We reached the Greenwich entrance in a matter of seconds, but Tad was forced to slam on the brakes. The gates we passed through only seconds earlier were now closed, chained and padlocked.

At that point, I grew a little more nervous and began to feel some serious trepidation about our situation. I summoned enough false bravado to suggest to Tad that he wait in the car with the engine running while I checked the chain and padlock.

If the lock wasn't completely fastened, I could open the gates, and we could leave the way we came. Tad curtly informed me that if I got out of the car I would be walking home because he was going to leave me there. Without further ado, he drove down the dirt road that runs parallel to the gates until he found a gap between two azalea bushes that was barely large enough for the car. Once free, we headed out the exit for the main road.

As soon as we left the cemetery, Tad pulled the car out in front of a Thunderbolt police car cruising the area, and we were stopped for questioning. Tad had left his driver's license in his car and we had just been caught leaving a cemetery that was closed. As you might expect, this suspicious behavior made for some interesting conversation with the police officer.

Tad concocted a quick lie to explain why he was driving my car with no license. He claimed to be test driving it while he considered its purchase. The cop was a good sport and played along with our story. He checked my license and made sure I was the legal owner and driver of the car. He appeared to be ready to let us go but then he abruptly asked, "What the hell were you boys doing back in that cemetery in the middle of the night?"

Tad managed to stammer, "We heard the cemetery was haunted and wanted to see for ourselves, but we got scared and came back out!" which was sort of the truth.

The policeman stared at both of us for a moment. Then he laughed at us and told us to get out of his jurisdiction immediately, an order with which we gladly complied.

TELEPATHY

The last personal experience I will share is one that happened when I was about sixteen years old. It was the middle of summer, school was out, and my sister had gone to the beach. It was a typical, routine summer day. It quickly turned unusual when I got home after baseball practice.

As I stepped through our front door, one of the worst feelings I can remember struck me. Even as I called out a greeting, I already knew nobody was home. What unnerved me so much was the thought I could hear my sister calling my name and asking for help, but not with her audible voice. I could hear my sister calling to me in my mind. Not only was I immediately certain something was wrong, I knew it involved my sister, Kathryn. I ran to my aunt's house down the street. Aunt Doty was outside on her porch, so I asked through her screens, "Aunt Doty, where's Kathi?"

I can't even remember her first answer, but I knew she was being evasive so I insisted she tell me the truth. It was very unusual for me to be that assertive with an adult at my age at the time. I was raised to be respectful of my elders, and my aunt knew me well. We were both very surprised as I insisted she tell me the truth and by the fact I knew she was lying to me.

Something about the expression on my face convinced Aunt Doty to tell me the truth. She said my sister was in the hospital at that very moment having her stomach pumped. Some idiot had spiked her drink at the beach, and she had lapsed into unconsciousness. She was rushed to the emergency room, the doctors pumped her stomach, and my sister would be fine.

I feel compelled to prove these events really happened to the best of my ability and were not simply products of my imagination. My wife slept right through the night when I met God, but I discussed the experience with her frequently in the following days. Jim Nettles may confirm the basic facts and that I'm not making up the ghost stories about his parents' house. My aunt could verify that I didn't invent the story about my sister and her overdose scare, and hospital records can confirm her admission and treatment. I can't offer conclusive proof of God or the supernatural by scientific method, but I can make every effort to soothe any doubts as to whether or not I believe these events to be true.

Because a news crew and team of scientists did not happen to be available when these experiences took place, does that mean they are not real. I want you to consider the possibility these experiences that I have shared with you are real. Our life views are shaped by our individual experiences. These come to comprise "truth" for each of us. Every story I recounted to you in this chapter is as real as the inexplicable experiences each of you also hold as true. There is simply more to this world than science can ever chronicle and quantify.

7 IS IT IN MY HEAD?

As I became immersed in the effort to write this book, I realized the crux of the debate would be a competition of my opinions against those of Richard Dawkins. Both of us have drawn on the work of others to form our opinions, and those people have been identified and acknowledged for their contributions. The scientific evidence, which exists, belongs to both Dawkins and I equally to claim and defend as best we can. I believe my arguments have been logical and very compelling up to this point.

We have evidence in astronomic observation of an expanding universe that essentially validates the Big Bang Theory. Natural selection has been observed first by Darwin on his trip to the Galapagos Islands and validated by over thirty years of research on the same islands by Drs. Peter and Rosemary Grant. The only other evidence for evolution is the fossil record, which allegedly provides proof of speciation.

Earlier we discussed why the fossil record does not prove evolution is true, but only that animals which are now extinct once roamed the earth. I demonstrated why the Big Bang still required God to speak before anything happened and without God there was no matter to explode. When the universe did not exist, there was nothingness; something had to account for the origin of matter. God provides the only logical answer, leaving us to ask "Where did God come from?"

Matter had to come from somewhere. In his book *The Creator and the Cosmos*, Dr. Hugh Ross suggested other dimensions exist besides the three we recognize using height, width, and depth measurements. Perhaps God came from these other dimensions. Perhaps heaven is in one of these extra dimensions. If God has no beginning but the Big Bang signals a beginning, only God could have created the matter on which the Big Bang relied.

Nevertheless, I realize the best outcome I could hope for would be a stalemate without my personal testimony based on my own first hand experiences being deemed admissible evidence. But I didn't write this book only to have this argument declared a draw. Anyway, Dawkins would win on points as being more credible since he's well known.

HELP WANTED: ARBITRATOR

I want to win. Ties are akin to kissing your sister. Therefore, I need an impartial third party to arbitrate this difference of opinion here and now. Yes, I have more to offer as evidence but, quite frankly, if my personal testimony is dismissed, the remaining evidence I wish to present should not be considered any more convincing. After all, eyewitness testimony is always given the most credibility in a court of law.

As I prepared for this debate, virtually every book I read was about evolution. I had previously read several authors sharing (and shaping) my system of beliefs such as Dr. Hugh Ross and Lee Strobel, so I refrained from reading other authors who supported my point of view. Instead, I read books by Tim Berra, Daniel Dennett, Bertrand Russell, and a few others to update my knowledge of evolutionary theory and evolutionist objections to Creation theory.

So my reading all fell into the two basic schools of thought, evolution or creation. I needed someone or a source of information to break the tie.

Having read Strobel's books in the past, I knew his personal story was both interesting and relevant. Given the fact that a significant part of Dawkins' objection to religion is his belief that children are indoctrinated into their faith at an early age and arguably against their will, Strobel disproves that theory. Dawkins believes that is the main reason adults cannot break free of their "fear of God."

Lee Strobel was never religious as a child and a staunch atheist as an adult. Strobel was an award winning legal editor for the Chicago Tribune. His investigative talents were put to good use as he set out to prove Christ was a myth. His wife had become a Christian, and Strobel felt her new faith threatened their marriage.

Strobel embarked on a two-year effort to find evidence that Christ never lived or if he did, he was not the Son of God. He feared his wife's "superstitions" and newfound faith in a false god would lead her to abandon him and their children and break up their marriage of twenty years. He became angry, as he grew more and more desperate to prove her religion was a lie.

When every effort to disprove the Bible failed, Strobel finally surrendered his life to Christ. He wrote a book based on his research called *The Case for Christ,* one of several books Strobel has written defending his new faith in God.

But Strobel falls into the creationist camp so he cannot arbitrate the stalemate to our debate. Naming him the arbitrator would be tantamount to stacking the deck, or cheating. Resolving this sticking point of whether Dawkins is justified in calling me delusional demands the selection of an impartial arbitrator with exemplary credentials.

I decided on the great philosopher and psychiatrist Carl Jung as the best man for the job. While I was not very familiar with his work beforehand, I recognized his name immediately as one of two iconic psychiatrists. The decision to choose Jung over Sigmund Freud was easy; Jung had written a book called *Psychology and Religion* that sat on the shelf near *The God Delusion* in my public library. Freud didn't have any books on the shelves nearby, so he must not have written a book dealing with the psychology of religion or it must have been checked out that day.

Jung was a sound choice as arbitrator. If my belief in God means that I am delusional, who would be better to say for sure than one of the most famous psychiatrists in the world? He certainly is more qualified to judge me insane than a zoologist. Jung's title suggested the book would answer my question – is it in my head?

The only problem was, his answer could well be yes, and that would really cut out work for me. Only one thing was sure; Jung would not object to settling our disagreement. He could not refuse my request for help since he's been dead more than 40 years.

If my religious belief is truly delusional as Richard Dawkins insists, the question that must be asked is this: why do so many people suffer the same delusion? It isn't just in my head, but in the heads of millions if not billions of others as well. Dawkins helpfully provided the definition of delusion as a "fixed, false belief." My dictionary varied only slightly: "a fixed, false belief that is resistant to reason or confrontation with actual fact (as in a paranoid delusion)."

Mental illness is commonly associated with being delusional. The word delusional is not complimentary. It's a very provocative word, one designed to provoke anger instead of thought. It's an insult intended to intimidate religious people or to provoke an angry response. The insult allows the very person who started the fracas by calling names to play victim and avoid further reasonable discourse.

DR. CARL JUNG ON PSYCHOLOGY AND RELIGION

"God is dead" – Nietzsche.
"Nietzsche is dead" – God.

The first line is a famous quote by the German philosopher, Nietzsche. The second line is the widely circulated joke that became popular after Nietzsche died. Dawkins repeatedly says belief in God is merely the product of a vivid imagination. I contend that religious experiences are real, not imaginary. To prevent an endless circle of accusation and rebuttal, I called on Dr. Jung for a third perspective. It was derived from the Terry Lectures Jung delivered in 1938 at Yale University. It should be noted; Jung identified the paradigms of introvert and extrovert from his work in his career and pioneered the study of the unconscious mind.

My expectations for support of my perspective were not very high. I suspected psychiatrists and psychologists would favor Dawkins' side of the debate due to the scientific nature of their medical training. I was prepared to read where Jung agreed religious people are weak-minded and less intelligent than the rational thinker. I braced myself to encounter a prejudiced opinion that accepted any evidence deemed scientific in nature and excluded all else.

In the very first sentence of his book *Psychology and Religion*, Jung refers to the "problem of religion." The phrase raised my concerns further. That exact same phrase was used at least twice more in the next few pages by Jung. I began to believe my worst fears were justified and I believed I'd be arguing against both Dawkins and Jung before I finished writing this chapter. In my vocabulary, the word "problem" has strong negative connotations. However, it turned out my fears were completely unfounded.

To have formed an opinion and quit reading after a few uncomfortable pages would never be sufficient. Never turn away from potential knowledge because you fear it will challenge your beliefs. I continued to read and was soon rewarded with a more positive outlook for my perspective.

By the end of Jung's book, I was completely engrossed and fascinated by his perspective. I half expected to be bored to tears by psychological mumbo jumbo, but found myself captivated by his logic and reasoning and blazed right through to the end of the book. The Terry lectures covered quite a bit of territory, so let's focus on the most interesting and relevant highlights as they pertain to our debate.

I always had assumed Nietzsche was an atheist because of his famous quote. Dr. Jung corrected my mistaken perception about the German philosopher. He explained what Nietzsche meant when he wrote God is dead–Nietzsche wasn't referring to the Judeo/Christian God called Yahweh or the God Allah by the Muslim, but meant the German god Wotan, also known as Dionysus.

Jung wrote "In Nietzsche's biography you will find irrefutable proofs that the god he originally meant was really Wotan, but, being a philologist and living in the seventies and eighties of the nineteenth century, he called him Dionysos." [39]

When reading anything written by Jung or Dawkins it's important to have a dictionary nearby. A philologist is a person who studies literary texts and written records. It seems the more educated the author, the more frequently the reader will need to look up a few definitions in the dictionary. Symptoms of advanced education include the development of an extended vocabulary and a burning, uncontrollable desire to show it off. Thank you, Daniel Webster.

Jung said, "The tragedy of Zarathustra is that, because his God died, Nietzsche himself became a god; and this happened because he was no atheist. He was too positive a nature to content himself with a negative creed…if he declares the "tremendum" to be dead, then he should find out at once where this considerable energy, which was once invested in an existence as great as God, has disappeared to. It might reappear under another name, it might call itself 'Wotan' or 'State' or something ending with–ism, even atheism, of which people believe, hope and expect just as much as they formerly did of God." [40]

So Nietzsche didn't believe in the Christian Creator/God, he believed in a pagan god that failed to live up to his expectations, so he replaced God with himself. Not only do I get it, I suddenly see parallels between Nietzsche and the thoughts of Bertrand Russell and Richard Dawkins. They all share in common the idea that nothing in the universe could surpass their own consciousness in terms of its greatness, suffering from ego morphed into hubris. Jung saw atheism filling the void left when God is driven from society through critical thought.

Reading something and comprehending its meaning can be two different things. More than once I became lost reading Jung, trying to grasp the meaning of an indecipherable word or a phrase that didn't translate into English. Other times I would slow down to contemplate a point Jung made about another religion and would find myself pondering the ramifications of that new knowledge added to what I already knew about that faith.

For example, Jung wrote that the Catholic religion teaches that Jesus' mother Mary ascended to heaven while still alive and took her earthly body with her. I was unaware of that tenet of Catholic faith despite the fact most of my extended family is Catholic. It never came up as a topic of conversation during any occasions spent with my Catholic friends and family. The idea required some time to consider the possibility and significance of such a thought as well as to reconcile what I believe about Catholics with what Jung was telling me.

I was familiar enough with the Catholic faith to have decided it was not for me personally before this information came to light and my opinion has not been altered. I do not believe in praying to saints or relying on intercession by others on my behalf to God. I don't see the need for including a priest as a middle man between myself and God. I prefer the "direct route" of approaching God personally.

After all, that's how I came to really believe in Him. This does not mean I think Catholics are going to hell for not believing as I believe or that my beliefs are all correct, and the Catholics are all wrong. It is not my place to judge the relationship between the Catholic and his or her Creator. It is enough simply to say that's not for me.

Jung does not speak of God as a trinity (a major bone of contention for Dawkins, by the way) but instead he refers to the "quaternity" comprised of four facets of God. Jung proclaimed the quaternal God to be Christian and the God of Gnostic philosophy as well. Initially this idea mystified me. The triune God is a concept difficult enough to understand without adding a fourth aspect to Him.

Christians know of the Holy Trinity as God the Father (Yahweh), God the Son (Jesus), and God the Holy Spirit (to which Christ referred as "the Helper"). Jung described the fourth facet of God as 'The God Within'. Jung suggested that putting God inside man creates the risk of limiting God's existence to the mind of the believer. We might conclude his concern is valid after also reading *The God Delusion* as Dawkins spends four hundred pages of paper doing his best to limit God.

Jung addressed this risk with a logical approach: "I have to admit the fact that the unconscious mind is capable at times of assuming an intelligence and purposeness which are superior to conscious insight. There is hardly any doubt that this fact is a basic religious phenomena…I have often met with the objection that the thoughts which the voice represents are no more than the thoughts of the individual himself. That may be, but I would call a thought my own when I have thought it, as I would call my money my own when I have earned or acquired it in a conscious and legitimate way. If someone gives me money as a present, then I will certainly not say to my benefactor, 'Thank you for my own money'." [41]

With a little humor and brilliant logic, Jung dismisses the argument that religious people are universally delusional. He lauded the power of the subconscious mind and acknowledged the superiority it holds over the conscious mind in perception of input outside the observable. The still, quiet voice inside each of us that reminds us of the difference between right and wrong is not Richard Dawkins' "meme" which we must now accept in lieu of divine inspiration. It comes from a more eternal source.

Jung also said, "I do not enjoy philosophical arguments that amuse through their own complications." [42] This confirmed my suspicion Jung would be glad to be dead, if he only knew he'd been drafted to be our arbitrator. But maybe Jung does know and is chuckling at his predicament.

Jung most certainly would not appreciate the mental calisthenics one must perform trying to follow Dawkins' logic. His logic is based on personal opinion steeped in bias against any consideration of God's existence but completely receptive to the possibility of spontaneous generation or self-initiated life. Dawkins excludes and ignores a wealth of information to reach his conclusions.

It doesn't matter to Dawkins that much of evolutionary theory cannot be proven by the scientific method any more than Creation science. He tells us Evolution science can be proven if just given enough time, and that Christian faith and proof are by definition mutually exclusive. Which will come first, proof of abiogenesis or the Second Coming? Evolution and Creation Theory both must use a "wait and see" approach to proof.

Richard Dawkins openly mocks and ridicules people of faith but gets his feelings hurt when he draws a reaction. It doesn't matter to him that faith in God is a personal asset to billions of people and has been for thousands of years. What matters to Dawkins is he retains the right to freely lambast the religious in the most derogatory of terms with impunity.

Dawkins asserts Christian faith is as misguided as belief in a creature he invented called a flying spaghetti monster. Simultaneously Dawkins refuses to dismiss the theory that we all live in a giant computer simulation model right out the plot of *The Matrix*, saying "we can't completely disprove it." Thanks for conceding the point, Richard, because we can't disprove God, either.

Jung made the point "Morality seems to be a gift like intelligence. You cannot pump it into a system where it is not indigenous, though you may spoil it. Unfortunately, there is no doubt about the fact that man is, as a whole, less good than he imagines himself or wants to be. Everyone carries a shadow, and the less it is embodied in the individual's conscious life, the blacker and denser it is." [43]

Dawkins would argue the point with Jung given half a chance. But after claiming atheists can be as moral as religious people, he suggests adultery is acceptable as long as the man loves both women and the relationships aren't based purely on lust. Morality is explained as the Darwinian return of favors in anticipation of payback known as reciprocal altruism that originate as "memes" inherited by our consciousness.

Dawkins claims man can be equally moral without believing in God as with Him although evidence to the contrary exists in abundance. Dawkins goes too far when he suggests Hitler was Christian. We have already established that atrocities have been committed by evil people who claimed to believe in God just as altruisms have been performed by atheists. The debate is larger than any one individual.

THE HUAORANI

The story of the first missionaries to the Auca tribe in South America serves as an example of what Jung meant. In the mid 1950s, this tribe of primitive Indians initially pretended to welcome overtures from missionaries, but without warning brutally murdered the men. That ended the story for the five missionaries, but not the tribe.

What happened afterward can only be described as either a remarkable set of coincidences or divine intervention. The Auca were notorious for behaving as savages, attacking neighboring villages and kidnapping their women. One kidnap victim escaped. Before leaving captivity, she learned to speak the Auca language and could communicate fluently in Auca. She met two women missionaries and taught them to speak the native's language. The two women missionaries evangelized to the tribe by communicating in their language after a long period of time spent gradually earning their trust.

The tribe responded so favorably to the message they all converted to Christianity. Their transformation was so total and complete that the tribe even chose to change its name. The Auca became the Huaorani, which translates as 'human' in their language. These people changed so completely they became evangelists themselves and now visit the same neighboring villages they previously terrorized to bring a message of hope and peace.

Amazingly, some anthropologists take a negative view of this reformation. They call it cultural imperialism. Apparently, they find it preferable these people should remain murdering savages rather than discovering God.

I attended a Steven Curtis Chapman concert a few years ago. During a break between sets, the story of the five missionaries and the Auca were told. Chapman introduced a member of the tribe who was identified as one of the killers of the five missionaries. This man who participated in the murder in cold blood of those missionaries years ago now atoned by serving as a dedicated missionary himself.

One last note to demonstrate the power of faith in God, and how it allows people to overcome the worst adversity–the two missionary women I mentioned who converted the entire tribe to Christianity? They were both relatives of the five murdered men. Their faith in God overpowered their fear of death and they literally followed in the footsteps of their slain loved ones.

It is truly incredible that people seriously debate whether the Auca were better off before they became the Huaorani, back when they still killed innocent strangers. Losing their "culture" of savage behavior to become what most people would consider role models is questioned as an improvement. Of course, it's easy to call it wrong and say it is cultural imperialism—from a safe distance. I'd love to know whether the rest of the jungle would prefer the Auca to the Huaorani for neighbors. Perhaps they can do a scientific poll and answer the question.

Even though he adamantly insisted he was not a religious man himself, Jung expressed fear of a society without God. He said, "If dull people lose the idea of God, nothing happens–not immediately and personally at least. But socially the masses begin to breed mental epidemics, of which we have now a fair number." [44] Jung recognized the shift in public attitudes contemporary to his times in apparent anticipation of the atrocities to come during World War II. These lectures were delivered shortly before its outbreak in Europe.

Jung established himself as an impartial observer, not a religious advocate of any sort. He said, "It would be a regrettable mistake if anybody should understand my observations to be a kind of a proof of the existence of God.

They prove only the existence of the archetypal image of the Deity, which to my mind is the most we can psychologically say about God." [45]

That reminded me somewhat of a personal experience. Several years ago a fellow co-worker took an intense dislike to me and began to campaign for our boss to fire me. For a while, it seemed her efforts would work. There were only three employees under our manager. There was me, another developer who worked with me with whom I was on cordial terms, and this woman who irrationally despised me. I asked my "friend" whose side he would take in our conflict. He just laughed and said to my disappointment, "I'm Switzerland," meaning he'd remain neutral in the argument.

Jung aspired to be "Switzerland" in this debate between atheist and believer and worked hard to demonstrate neutrality. By the end of his book it was still easy to conclude Jung was predisposed to favor belief in a God and surprisingly critical of atheism. His predominant message was religion cannot be rationalized or otherwise taken away from the believer who has a personal relationship with God.

Having personally experienced something of that very nature, someone could tell me a million times it was only a hallucination, and I still would believe my experience really happened. It's much easier to believe because it happened to me. If someone else told me of a similar experience, it might impress me profoundly, but it wouldn't carry the same weight with me as having my own personal experience did. Any attempt by someone else to discredit, diminish, or rationalize my experiences in an attempt to explain what happened by someone else, I will reject because they didn't share the experience with me. They're guessing what happened and are trying to explain my own experience to me.

Because I was there, I think I am a better judge than someone who wasn't sharing the experience with me and therefore could not possibly know what really happened. You don't have to believe me. I am willing to take a polygraph test (at the requestor's expense, of course); if it will help the skeptics accept at least my contention that I honestly believe my story. I didn't invent it for your amusement. It would be much easier for me to accept evolution as Dawkins would have me do.

But Jung expressed a prejudice against atheism with vehemence that surprised me: "The philosophy of the Upanishads corresponds to a psychology that long ago recognized the relativity of the gods. This is not to be confounded with such a stupid error as atheism." [46] Jung summarized his lectures with the following stunning rebuke of the biased atheist.

He described patients to his practice who have fully embraced their religious beliefs and need those beliefs in order to be happy. By making their choice between "the devil and the deep blue sea" where the devil represents the "Mandela" symbol for evil, and the deep blue sea represents the comfort of the religious "neurosis," Jung said those patients led happier and more fulfilled lives than they would have without choosing to be good.

He rebuked the atheist and said, "The well-meaning rationalist will point that I am driving out the devil by Baalzebub and that I replace an honest neurosis with the cheat of a religious belief. Concerning the former, I have nothing to reply, being no metaphysical expert, but concerning the latter, I must point out there is no question of belief, but of experience. Religious experience is absolute. It is indisputable. You can only say you have never had such an experience, and your opponent will say "Sorry, I have." And there your discussion will come to an end." [47]

That was perfect. The religious man may not convince the atheist of the merit in his faith, but he doesn't have to. The atheist cannot take the believer's faith away from him any more than the believer can give his faith to the atheist. Both must be willing partners in the exchange.

Jung continued, "No matter what the world thinks about religious experience, the one who has it possesses the great treasure of a thing that has provided him with a source of life, meaning and beauty and that has given a new splendor to the world and to mankind. He has pistis and peace. Where is the criterium by which you could say that such a life is not legitimate, that such experience is not valid and that such pistis is mere illusion? Is there, as a matter of fact, any better truth about ultimate things than the one that helps you to live?...the thing that cures a neurosis must be as convincing as the neurosis; and since the latter is only too real, the helpful experience must be of equal reality. It must be a very real illusion, if you want to put it pessimistically. But what is the difference between a real illusion and a healing religious experience? It is merely a difference in words." [48]

Carl Jung probably never knew Richard Dawkins, but this next quote could have been specifically written about him: "You can say, for instance, that life is a disease with a very bad prognosis, it lingers on for years to end with death; or that normality is a generally prevailing constitutional defect; or that man is an animal with a fatally overgrown brain. This kind of thinking is the prerogative of habitual grumblers with bad digestions. Nobody can know what the ultimate things are. We must, therefore, take them as we experience them. And if such experience helps to make your life healthier, more beautiful, more complete and more satisfactory to yourself and those you love, you may safely say: This was the grace of God." [49]

Amen.

8 SPOOKY STUFF

On the night of December 11, 1978, fifteen year old Robert Piest was about to get off work for the night from the local pharmacy where he was employed. His mother had arrived to give him a ride home; Robert grabbed his coat he'd lent a co-worker and asked his mother to wait while he walked outside to the parking lot to speak with a man about a job doing construction work. Robert was never seen alive again. [50]

Des Plaines police lieutenant Joseph Kozenczak quickly became frustrated by Robert's disappearance and the lack of clues. Robert was popular in school. For a teen with his profile to go missing was strange. The manner in which he disappeared made the case even stranger–he vanished from the parking lot of the store where he worked while his mother was there to take him home and their family waited for their return so all could celebrate Mom's birthday. These details bothered the detective and left him with an uneasy feeling that the case involved more than a teen who had run away from home.

PSYCHE IT

Eventually in desperation, Detective Kozenczak decided to consult with a local psychic named Carol Broman for help with the case. He met her early one morning at her home with a few of Robert's possessions available in order to help strengthen her psychic impressions. Detective Kozenczak was completely unprepared for the nature and the level of detail in the information Broman would give him.

Broman told him not only was the boy dead, but there were a number of other victims buried near his body. Broman told the police they weren't looking for a murderer—they were hunting a serial killer. She had disturbing visions of the victims being tortured, sexually assaulted, and choked to death using a rope with a dowel to help twist it tighter.

The information that Carol Broman provided led the police to John Wayne Gacy. She had told police that the killer was about forty years old, had a scar on his little finger and walked with a limp. That perfectly matched the description of Gacy right down to the scar. On December 20, 1978, nine days after Robert's disappearance, Gacy was arrested on a drug bust. The arrest gave the police probable cause. They had Gacy under surveillance based in part on the information provided by Broman. A search warrant was issued for Gacy's residence in Des Plaines. There police discovered twenty-nine bodies buried in the crawl space under his house. A claim ticket the coworker who borrowed Robert's coat had put in its pocket was also found in the house. With that clue, the police confirmed Robert Piest was another of his victims.

Information from Gacy himself led the police to another four bodies in the surrounding area, but Robert's body was not among them. Gacy told police he had thrown Piest's body from his car off a bridge about 65 miles outside of Chicago.

Robert Piest's mother knew the police had worked with Carol Broman to help find Robert's killer. Elizabeth Piest had read an article in Reader's Digest magazine where she learned about the work of another psychic named Dorothy Allison who lived in New Jersey. She asked Lieutenant Kozenczak to contact Dorothy to ask for her help in recovering Robert's body.

Even though she had no personal knowledge of the Chicago area or Des Plaines, over the phone Dorothy was able to describe in detail the location where Robert's body would be found. She predicted the smell of diesel fuel would be a prominent "tell" when they drew near his body and there would also be a small cemetery called Evergreen in the vicinity. Dorothy made clear to the detective during their phone call that these landmarks would be found as features of the surrounding landscape near the location of Robert's body.

On request, Dorothy flew in from New Jersey and rode out into the countryside with the detective. She directed Kozenczak past some old diesel storage tanks to a spot near an old abandoned cemetery–a cemetery detective Kozenczak later discovered had been named Evergreen.

Dorothy pointed to a bend in the frozen river. She told the detective that's where Robert's body would be recovered when the ice thawed and melted. Before leaving to return home, Dorothy scribbled a date on a slip of paper: April 9, 1979. Dorothy gave it to detective Kozenczak and told him the date written on the note would be the day that Robert's body would be recovered from the river. Kozenczak tossed the note inside Robert's file and returned it to his drawer, barely glancing at it. He was still disappointed they had been unable to recover Robert's body immediately.

On April 9, 1979, the sheriff of nearby Grundy County called Detective Kozenczak to inform him that Robert's body had just been recovered from the melted river. The detective remembered the note and checked Robert's file. The note confirmed Dorothy had accurately predicted the exact date he would be found months in advance. When Detective Kozenczak met the sheriff and was shown the location where the body was recovered, he realized it was only a few hundred yards from where Dorothy Allison had pointed several months earlier.

In solving the case of serial killer John Wayne Gacy, two different psychics provided police with critical information that led to the arrest of one of the worst serial killers in American history as well as the recovery of the body of the victim who directly led to his capture. There are only two possible logical conclusions one can make about these two psychics, given the detail of information they provided during the investigation. Either both women are frauds and somehow they were both accomplices of Gacy and therefore privy to details of the crime(s), or they possessed real abilities, which were put to good use and stopped an evil murderer in his tracks. Considering Dorothy Allison's help had been solicited and not volunteered, that would make the coincidence extremely implausible. But this is only one example.

TENNESSEE CASE

In Memphis, Tennessee police arrested a man named Larry Ward for the murders of two prostitutes. Psychic Carol Pate voluntarily contacted homicide detective Ken East, the lead investigator in the case. She told him he had arrested the wrong man. Detective East thanked Ms. Pate for her offer of help, but several witnesses had identified Ward as the killer and quite frankly, he didn't see how her "assistance" would strengthen his case for the arrest which was already made.

When a third prostitute was murdered, Ward had an iron-clad alibi. This meant, in all likelihood, the police had the wrong man as their prime suspect. Detective East conducted follow up interviews with the two witnesses that identified Ward as the gunman in the first two murders. He quickly found out that both witnesses had held a grudge against Ward and both lied to implicate him. The case was back at square one with no real leads or suspects.

The detective remembered Carol Pate's offer. He called to ask if she could meet with him at one of the crime scenes. He hoped she would be able to provide additional information about the killer if she physically visited the location. When they met at the scene of the second murder, Pate asked for a photo of the crime scene that included the victim. Looking at the photo, she described the murderer as a black man of medium height who was roughly forty years old and had a goatee.

Pate said the killer had multiple personalities. She identified the killing personality as an evil spirit named Al. Before their meeting ended, Pate told Detective East that there would be more murders before Al was caught.

True to Pate's prediction, the killer struck again, murdering another female prostitute. But then police caught a second break in the case. A transvestite named Tony Thompson was shot twice but survived and was able to give the police a composite sketch of his attacker, one that matched the detailed description Pate provided earlier.

Pate contacted the police again. Somehow she had seen into the future and knew that Al was about to strike again. This time the murderer would attack his victim near a church with a stone front of which there were only a few in the area where Al hunted. Police beefed up their patrols in the area, which put an officer near the scene when the next shooting occurred.

Thompson had vaguely described a dark grey sedan driven by the killer. The policeman was able to get the tag number of a suspicious vehicle matching Thompson's general description of his attacker's car. The tag number matched a vehicle belonging to James Barnes, a local deliveryman.

Police interviewed Barnes' wife and decided he was a credible suspect, but still lacked sufficient evidence or probable cause to put him under arrest or charge him with any of the crimes. Instead, Barnes was kept under surveillance until he resumed prowling his hunting grounds and gave police probable cause. They took Barnes to headquarters for questioning.

During questioning, Barnes became agitated and he asked to be handcuffed. The investigators conducting his interview thought the request very odd. When they asked Barnes why he wanted to be handcuffed and referred to him by name he said, "My name is not James Barnes." He insisted his name was Al.

The officers later learned Barnes had begun killing the prostitutes in an attempt to placate Al, who wanted Barnes to murder his wife and children. Not only had Pate accurately told police they originally arrested the wrong man, she correctly identified the killer as a man suffering from multiple personality disorder. She even correctly named the killing personality as "Al." She provided police with a detailed description that perfectly matched a composite drawing made with the help of an actual survivor. Police handling the case have remarked that witnesses who actually see a crime in progress do not usually provide information as solid and accurately as Pate. From wherever she got her information, it was better than first hand and as reliable as the very best eyewitness. [51]

After working on the case Detective East said the experience made him a believer in psychics. But not every investigation involving psychic assistance is an unsolved murder or involves a serial killer; sometimes stories of psychics helping the police even have happy endings.

AGAIN IN NEW YORK

In 1975, a six year old boy named Tommy Kennedy wandered away from his family's campsite during a thunderstorm in upstate New York. Over two hundred people quickly formed a search team and frantically combed the woods. It was a race against time and the elements to find the young boy while he was still alive, before exposure, a wild animal, or some accident befell him.

The sheriff had heard of Phil Jordan, a local man who supposedly showed some psychic abilities. He called Phil and asked him to join the search. Jordan agreed to help but refused to come immediately, insisting he needed to wait until morning to have time to prepare. The sheriff was upset and disappointed that Jordan wouldn't come immediately, but he felt he had no choice but wait for him. Darkness had fallen with no luck finding young Tommy.

When Jordan arrived the following morning, he brought with him a detailed map of the area he had drawn overnight. Jordan claimed while meditating he had seen the boy's location in a vision. The sheriff handed Jordan one of the boy's sneakers, as psychics often want contact with physical objects belonging to the victim in order to sharpen their intuitions.

Jordan's map accurately depicted a lake near the campsite and a clearing beyond the lake. He said the boy could be found near the clearing on the far side of the lake's shore. Most of the searchers found it hard to believe the boy would have safely gotten as far as Jordan's map pictured and continued to scour different search areas. A small team accompanied Jordan as he followed his hand drawn map.

Jordan's team had gone several miles and ventured just far enough that the enthusiasm for following his lead waned. Doubts among the searchers grew. Members of Jordan's group groused and speculated they were on a wild goose chase. When they expressed their doubts audibly, Jordan even began to doubt himself for a moment. Then they reached another landmark on the map, and he realized they were very close. Minutes later, the boy was found safe and sound.

Televised interviews of the sheriff, Phil Jordan, and Tommy Kennedy were rather straightforward accounts of the same story told with complete sincerity. Each version of the account corroborated another. Tommy was too young to be a willing accomplice. He had no reason to help Jordan, if he had actually been kidnapped so the event could be staged.

Years later, video of an adult Tom Kennedy showed a man who seemed appropriately grateful to Phil Jordan for helping save his life. There hadn't been time for a "Stockholm syndrome" to take place where the victim became friendly to his kidnapper. In short, it all appeared as if the story had happened exactly the way it was told. [52]

There is sufficient documentation proving the veracity of the statements surrounding the search. Jordan allegedly had never visited the area where Tommy went missing prior to arriving to help search. He had no reason to be familiar with the lake or other landscape features, but Phil came prepared with a detailed map drawn with his own psychic vision of unfamiliar terrain. His visionary map led to the dramatic rescue of a six-year-old boy lost deep in the woods.

Like Dorothy Allison, Jordan did not volunteer his services. He was sought out by the sheriff. Like the Gacy and Barnes murder stories, Tommy Kennedy's story sounds impossible to believe. Yet police officers involved gave taped interviews confirming the facts and details of these cases.

There are hundreds if not thousands of cases like these. Reports abound of psychics and people with abilities that defy logic. While I would never choose to consult a palm reader or a psychic under normal circumstances, sometimes desperate times call for desperate measures. If one of my children or grandchildren were missing like Tommy Kennedy, how can I really say to whom I would turn to for help? I suspect I would be open to help from any and every available source, unless I was certain the source was evil.

Carol Pate, Carol Broman, Dorothy Allison, and Phil Jordan share the common trait of psychic ability, but they are special because they use their gifts to help others. They don't strike me as evil people, but good people with a gift they put to good use. They are contemporary examples of psychics or clairvoyants with unexplainable and remarkable powers. These people are not famous household names. However, there is one name with whom most people are familiar. Nostradamus.

NOSTRADAMUS

Nostradamus lived from 1503–1566 and became famous during his own lifetime for his books that foretold the future. Nostradamus was famous for writing in the form of quatrains that disguised his "visions." Some have said this was done to avoid reprisals from the subjects of his predictions. While Nostradamus achieved fame in his lifetime; it was nothing compared to the fame he enjoys posthumously today.

There have been television shows, books and websites devoted to revealing and explaining Nostradamus' work in detail. One even starred the late actor Orson Welles, himself an avid believer in Nostradamus. Many believe Nostradamus accurately predicted the deaths of kings, the outbreak of wars, and some even believe he foretold of the terrorist attacks in New York on September 11, 2001.

Because his quatrains were cryptic in form, it is somewhat difficult to say for sure which of his predictions may or may not have come true. Nostradamus did not specifically name any people or events with pinpoint accuracy, but he did give some interesting clues that may have foretold of future events. The closest he came to specifically naming a leader and a country involved in a war was when Nostradamus wrote of Hister in Germany, which many have assumed really meant Hitler.

Nostradamus was an occultist who studied astrology and occult texts, but he was also a practicing Roman Catholic. Nostradamus strongly believed that being a Christian and belonging to the Church was essential to one's salvation, but he dabbled in what most consider "black arts" as well.

EDGAR CAYCE

Nostradamus could be called a charlatan or a piker compared to Edgar Cayce. Nostradamus wrote in vague prose that could have multiple interpretations, but Edgar Cayce gave specific and detailed information in his readings. In the course of his lifetime, Edgar Cayce gave an estimated 30,000 psychic readings to clients including President Franklin Delano Roosevelt, Thomas Edison, and John D. Rockefeller as well as Cayce's own family and neighbors in rural Kentucky. Over 14,000 of the readings were recorded by Cayce's stenographer. The detail and subject matter are fascinating.

Cayce never charged a penny no matter who the client; he relied on gifts in whatever amount the "entity" receiving the reading would offer. Cayce could give readings remotely; as long as he knew where his subject was, he could communicate with them. Cayce was called "The Sleeping Prophet" because his technique was to simply loosen his tie, take off his shoes, and take a nap. He would go into a trance similar to a sleep state and answer questions if the timing was right. If a question was put to Cayce just before he went completely to sleep as his eyelids began to flutter, amazing and incredible information could be extracted from a consciousness he was able to tap.

Somehow, a poorly educated former salesman from rural Kentucky was able to dictate in medical terms state-of-the-art surgical techniques. In fact, Cayce has been credited with pioneering orthopedic surgery when he recommended hammering nails through the man's kneecap into the broken bone to hold it in place while it healed.

His neighbor with a badly broken leg had been told he would never walk normally again. Rather than accept his prognosis, the neighbor chose to get a psychic medical reading from Cayce. The surgery was performed as Cayce described, and within weeks the man walked normally again. That is only one instance of Cayce providing a medical diagnosis and treatment plan superior to experts of his day. Cayce saved his wife's life from pneumonia after she was given weeks to live. Cayce was also credited with curing a nephew of blindness that was believed to be permanent, but radical treatment plans saved his wife's life and the boy's sight.

Cayce predicted major, world changing events with startling accuracy, including the Great Depression and World War II, both several years before they happened. Not only did he correctly predict the stock market crash, Cayce specifically named the month and year.

Unfortunately, Cayce didn't benefit from that knowledge and lost everything during the Depression, including the hospital he had dreamed of and built. He wasn't prepared for the Depression because he didn't know it was coming because remarkably, Cayce never remembered any details from his readings when he awakened. He simply didn't remember that he had predicted the disaster several years before.

Not only did Cayce predict the start of World War II, he named Japan, Austria and Germany as members of the same coalition that would start the war. The war was ultimately responsible for Cayce's death. He worked himself beyond exhaustion, giving between six and eight readings per day when he knew that two was his limit. He would not refuse helping people who wanted information about a loved one involved in the war. Cayce suffered a stroke and died on January 3, 1945. Ironically Cayce's last prediction was the date of his own death.[53]

So, what do these stories mean? Should we put all our trust in psychics and start consulting our horoscopes? Do these people represent contact with an alternate power that replaces or supplants God?

The Bible clearly says that we shouldn't dabble in the occult, but Cayce and many others with psychic abilities have professed a strong faith in God, so were they wrong to tap into these skills? These are but a few examples of people with abilities that allow access to information that "normal" people don't get.

We find ourselves trusting Phil Jordan, Carol Pate, Carol Broman, or Dorothy Allison because their intentions are good. Edgar Cayce struggled with the conflict between his religion and his psychic ability. His Christian faith in the Bible meant he was aware it forbade using mediums, but he also knew he had a powerful gift he could put to use to help other people. Cayce refused to use his gift for profit.

He only accepted gifts or donations from clients when he could have commanded a fortune, once his fame and reputation spread. Cayce seemed to successfully walk that fine line between using and abusing his phenomenal gift.

The point of including these stories and this chapter about psychics in a book debating the existence of God versus evolution as the sole explanation for life as we know it is this: open your mind to other possibilities outside the obvious. Beyond what we see, hear, feel, touch and smell. There is so little about this world that we do really know. Does believing in scientific theory require denial of the supernatural?

An older, wise physician once asked a colleague who scoffed at the idea that life after death was possible: "How much of what is true do you think we know? Five percent? Ten percent? Certainly, what we don't know is far greater than what we do know. Science is like having a very bright flashlight in a gigantic cave."[54]

Science cannot begin to explain how these people could do what they apparently were able to do. You can't have it both ways. Science can't dismiss the supernatural as superstition and nonsense without coming up with a viable explanation for how these things can happen.

Just because these stories are not subject to scientific method or experiment because they cannot be reproduced within a laboratory does not make them untrue. The nature and unpredictability of these phenomena make them difficult to study but, not unbelievable, especially to those who experience them first hand.

By the same token, if these psychics do have the skills and abilities described, how does that impact the idea that a Supreme Being could exist and was responsible for the information these people are tapping?

It would seem that belief in a supernatural God would be a natural extension from believing these phenomena are true.

The information these truly gifted psychics are tapping into has to come from somewhere. The most logical explanation is supernatural or psychic information comes from God in much the same way dreams and visions were sent to the prophets of old.

9 REASON TO BELIEVE

Angels beating all their wings in time
With smiles on their faces
And a gleam right in their eyes
Thought I heard one sigh for you
Come on up, come on up, now
Come on up, now

May the good Lord shine a light on you
Make every song you sing your favorite tune
May the good Lord shine a light on you
Warm like the evening sun

-Mick Jagger/Keith Richard [55]

In "Personal Experiences" I wrote about the night I believe I had a close personal encounter with God. I also shared what I personally saw and believe to this day to be the paranormal activity of ghosts; things I witnessed on a number of different occasions at the house of my friend Jim Nettles in Savannah, Georgia.

Then in "Spooky Stuff" I recounted stories of true and documented cases where psychics helped police solve horrible murders and stopped two serial killers. The police were given information they could not possibly have attained through normal channels. If psychics really can get information from sources we cannot see, hear or touch—if ghosts exist and my experience with God was real–could there be evidence available of any kind that serves as a sort of scientific proof that some supernatural phenomena is real?

Countless "near death" stories exist which, have been researched and backed by documentation that describes in eerily similar detail how many people have temporarily died and come back to life.

Sometimes they come back due to medical resuscitation. In other cases, the patient is revived somehow because of more inexplicable reasons, even after the subject was given up for dead. The strong, recurring common denominators in their stories were the documented medical facts of clinical death followed by a restoration and resumption of life. The similarities in accounts given by those who undergo the phenomena describe fantastic experiences while in a medically dead state. In many cases, these people experience changes that can only be described as beneficial or evolutionary in nature.

Melvin Morse was a Seattle pediatrician who had his life changed forever when he asked a few simple questions of a seven-year-old patient named Katie. Katie made a remarkable recovery after she nearly drowned in a community swimming pool accident. Katie remained comatose for three straight days with little hope for her survival.

Although she had been clinically dead for nineteen minutes, Katie's case was unusual because she showed no signs of permanent brain damage. Irreversible brain damage is normally expected when anoxia (a lack of oxygen to the brain) occurs and the heart stops beating for such a long period of time.

While treating her, Dr. Morse observed Katie's pupils were fixed and dilated during the three consecutive days she remained unconscious. This commonly indicates the patient is in an irreversible vegetative state. However, three days later, Katie suddenly made a full recovery.

Dr. Morse wanted to learn more about Katie's accident. He casually asked if she remembered anything about how she fell into the pool. He hoped Katie could recall enough of her experience to determine whether she slipped and hit her head or possibly had suffered an epileptic seizure. At the time, Dr. Morse was only interested in the cause and not the effect of Katie's close encounter with death. He was completely unprepared for her answer.

Katie told Dr. Morse she spent those three days visiting other worlds and places while her physical body remained in the hospital. To his astonishment, though in a comatose state, Katie was able to describe in detail which doctors had treated her and what procedures each had done as Morse and his team worked feverishly to save her life. Somehow, while her body lay unconscious in bed, she watched what happened around her.

Katie then told him of an angel who escorted her to her home on earth. There she watched her mother cook dinner and her little brother play with his toys as they prepared to return to the hospital. Katie finally described going through a long, dark tunnel toward a bright light where she met a being she called the Heavenly Father and spent time in heaven.

For a moment, Dr. Morse remained stunned into speechlessness. After all, her story defied everything he knew from his medical training and personal experience. A few moments later, he managed to ask Katie what it had been like "up there." Apparently the shock and disbelief was evident on his face as Katie answered, "Don't worry, Dr. Morse–heaven is fun!" [56]

As her personal doctor, Morse knew that Katie's experience defied conventional neurology. Her symptoms had indicated the absence of any brain function. She should not have been able to comprehend or remember anything after falling into the pool.

His medical knowledge and experience told him that her story was impossible, but it also told him it was impossible for her to be sitting in the room answering his questions with no apparent ill effect from her ordeal. Dr. Morse believed her. Part of the proof for him was her confident descriptions of details of her treatment during which she was not conscious. Even more convincing was her manner while telling her story. It was obvious Katie believed her experience was real.

Stories of this nature occur so frequently that medical professionals and scientists have begun collecting and studying data and recording the experiences of people who were medically declared dead and then resuscitated. Not every doctor or scientist believes these experiences really happen, but most do accept as true that their patients are convinced the experiences really occurred to them.

The pioneer of this field of research was Dr. Raymond Moody, who first collected stories of near death experiences from his patients and published them in a book called Life After Life. Moody actually coined the term "near death experience" to describe his new field of study and has interviewed thousands of patients and conducted the first attempt to study the topic seriously.

Moody was the first to notice common denominators in the NDE stories his patients told. He documented that true NDEs included several, if not all, of the following nine characteristics: a sense of being dead, feelings of peace and painlessness, an out-of-body experience, the sensation of traveling through a dark tunnel, meeting people of Light or dead family members, meeting a "Being of Light" or God, having a life review, a feeling of reluctance to return to this life, and having a personality transformation as a result of the experience.

After talking with Katie, Morse initially remained somewhat skeptical. He and his colleagues at Seattle Children's Hospital embarked on a small study of 26 children who had nearly died for various reasons and compared their stories with those of 151 other children who had been extremely sick but had not nearly died. The control group of children may have even been admitted into intensive care and required a ventilator for breathing, but never went into cardiac arrest.

Morse found that 23 of the 26 children who nearly died had an NDE and described several of the nine traits of an NDE identified by Moody, but none of the control group of children experienced anything similar. Results of those interviews became the inspiration for Dr. Morse's first book on this subject, Closer to the Light.

SCIENTIFIC STUDIES OF THE NEAR DEATH EXPERIENCE

Morse needed to know more. He designed and conducted what is now called the Transformations Study (or Seattle Study). It was intended to determine if any effect of the experience impacted the lives of adult patients who had an NDE in childhood.

Four hundred people participated in the Transformations Study divided into five groups. The study group consisted of adults who experienced a NDE as a child and a minimum of ten years had elapsed since the experience. The control groups were organized into the following categories: adults who were extremely sick as a child but did not have an NDE, adults who claimed to have had vivid experiences with a warm and loving light as either an adult or child but not as part of an NDE, adults identifying themselves as mystical or spiritual, and finally adults reporting an out of body experience (OBE) as a child.

Participants in the study groups were carefully matched. Age, sex, the medications administered during their treatments and any anesthesia used were all separate matching criteria. Study participants represented a cross section of society including doctors, lawyers, housewives, musicians, accountants, and others spread geographically throughout the U.S. The purpose of the study was not to determine whether or not the person claiming to have a NDE actually had one, but whether or not the alleged experience had any impact on their life.

The conclusions Morse drew from the Transformations Study were remarkable. Those who experienced an NDE as a child had a strongly diminished fear of death, an increase in psychic abilities, and a higher appreciation for life. Some even claimed to be smarter after the experience. While those claims are the most difficult to substantiate, it should be noted that one individual began doodling as a habit after his NDE. Later his "doodles" were discovered to actually be complex physics equations originally developed by Max Planck, the father of atomic theory.

Morse documented the transformations of the NDE group by creating a base line from the four control groups. Every participant was given an extensive battery of tests to complete. Each study participant was also interviewed by researchers for a minimum of four hours. Trick questions were included to check for imposters among the volunteers for the study.

Tests included two death anxiety scales used to evaluate the study participant's relative fear of death. The Templar Death Anxiety Scale is composed of 15 true/false questions like "I often think about how short life is" or "I am scared of having a heart attack" and was designed to measure the test taker's relative fear of death. The Reker-Peacock attitude survey assessed the subject's relative satisfaction with the quality of his or her life. Other tests were used to assess personal values of study participants and the validity of claims by some of having increased abilities resulting from their NDE.

The Greyson Values Survey was used on each study participant to identify what was personally important to the respondent. The Subjective Paranormal Events Questionnaire was given to see if the study participants had really developed psychic powers. Researchers participating in the study made a concerted effort to scientifically answer the question of whether people that claimed to have had a NDE really gained from the experience.

The emphatic conclusion of the study showed that these patients did change for the better and remarkably so. Only the OBE control group came close to scoring as highly as the NDE group on scales measuring quality, satisfaction, and enjoyment of life. Dr. Morse recorded the findings from the Transformations Study in his second book called Transformed by the Light. In addition to a number of interesting cases culled from the study, the book includes a number of fascinating NDE stories by people who were not included in the study due to its strict guidelines.

One of those was a particularly remarkable story not of a NDE, but an actual death that was claimed to be psychically shared by a sibling. One day a distraught mother contacted Dr. Morse to ask if a NDE could be shared by a living person. Her son recently died in a traffic accident after several hours of being kept alive on life support.

After he passed away, the boy's mother and father went home to break the news to his younger sister. She was hearing impaired and therefore could not be called and informed of what had happened. Yet when the parents arrived home and went to tell the boy's sister the tragic news, they found her talking aloud in an empty room. She told her shocked parents she was speaking with her dead brother.

The mostly deaf sister acknowledged that she knew her brother was dead. She said she had awakened from a nap after school to find his ghost with her. Not only could she see him, she somehow heard him speak. The sister claimed when they arrived she had been able to see both the room they were in and into another world.

The girl said her brother took her to heaven where she saw relatives who had previously died. She even told them her brother had told her a "secret." She claimed he told her their aunt was pregnant and would have a boy. This was unknown to the family at the time but later confirmed to be true.

Before including the story in his book, Dr. Morse verified the account of the boy's death. He interviewed the family members individually and spoke with character witnesses who vouched for the family's integrity. Through his research, Morse determined the story to be true. Although he could not prove the girl had actually been contacted by her dead brother, he was able to corroborate the details of the boy's death and aunt's pregnancy.

Dr. Morse concludes only two possible explanations could account for these NDE stories; either these stories are repetitive falsifications or something really happens to many patients when they experience death temporarily, and they are describing a reality for them. Dr. Morse has accepted that his patients really do experience what they claim or at least they sincerely believe they have, and he considers it a form of patient neglect or even abuse for the physician to argue against the patient's best interest or claim the event was only a delusion.

With help from a colleague, Dr. Morse documented cases of NDEs that occurred in extremely remote parts of Africa. Surely media exposure or coverage of this phenomenon is virtually non-existent there–most people who live there don't even have televisions. The possibility of contamination of those NDE accounts is virtually nonexistent.

However, details of the African NDE accounts were basically the same as others. The difference came in the perceptions by those who experienced the phenomena in part of the reactions of family and friends who tended to suspect the individual who recovered from near death was bewitched. Most African NDE patients did not view the experience as positive because of their negative treatment from society.

In Europe, doctors and scientists have also begun to study NDEs. A three-year study was conducted in England at Southampton University of 1500 heart attack survivors in hopes of determining whether patients without a heartbeat or brain activity can have an out-of-body experience. EEGs confirmed electrical activity (and therefore brain activity) ceased completely but eleven percent of those patients recalled emotions and visions during that time. Conclusions drawn from the study suggested consciousness may continue to exist after the brain ceases to function and the body is clinically dead.

On December 15, 2001 the medical journal, The Lancet, published results of a thirteen-year Dutch study that investigated NDEs reported from ten different Dutch hospitals. A number of patients who had suffered cardiac arrest and then were resuscitated were interviewed. The purpose of the study was to find out what percentage of patients experienced a NDE and its effect on their lives afterward.

According to The International Association for Near Death Studies (www.iands.org) "The advantage of this type of study is that it gives scientists a matched comparison group of non-NDE patients against which to compare the near-death experiencers, and that in turn gives scientists much more reliable data about the possible causes and consequences of the near-death experience."

Doctors conducted follow up interviews to find out whether or not the experience had transformed the participants and to what degree. Basically, they were looking for the same information pursued by Dr. Morse in his study but on a much larger scale. Like the Transformations study, the Dutch study found NDE survivors tended to be happier and led more fulfilled lives than the general population.

According to its chief investigator, cardiologist Pirn Van Lommel, "Our results show that medical factors cannot account for the occurrence of NDE. All patients had a cardiac arrest, and were clinically dead with unconsciousness resulting from insufficient blood supply to the brain. In those circumstances, the EEG (a measure of brain electrical activity) becomes flat, and if CPR is not started within 5-10 minutes, irreparable damage is done to the brain, and the patient will die. According to the theory that NDE is caused by anoxia, all patients in our study should have had a NDE, but only 18% reported having an NDE.... There is also a theory that NDE is caused psychologically, by the fear of death. But only a very small percentage of our patients said they had been afraid seconds before their cardiac arrest—it happened too suddenly for them to realize what was occurring. More patients than the frightened ones reported NDEs." [57]

While the Dutch study did not break new ground, it affirmed the results of the earlier, smaller study by Morse. The truth is millions of people around the world have claimed to have had a NDE; a Gallup poll conducted in 1992 estimated that 13 million Americans or five percent of the total population had been through a near death experience. [58]

Dr. Morse's book documented the first serious attempt to apply the principles of scientific method to the study of near death experiences. Earlier studies of NDEs involved only collecting anecdotal tales of individual experiences through an interview process. Interviews were only a small part of the effort for the Transformation study.

Morse realized it was important to document every detail in every case, from obtaining medical records describing to the treatments and drugs used to confirming with attending medical personnel that clinical death had occurred and been declared by them. Morse and his team conducted exhaustive interviews with each study participant and applied a full battery of tests as well to obtain measurable results from his research while following solid protocol for the scientific method. Morse never set out to prove any of the experiences were true, but only to study the effects a NDE had on the person who experienced it.

By no means is Morse's book the only one on the subject of NDEs; it isn't even his only book on the subject. Anecdotal accounts of NDEs have been recorded in a number of books including tomes by Morse and Moody. Most are by authors who describe first hand NDE experiences.

Embraced by the Light by Betty Eadie, Saved by the Light by Dannion Brinkley, and both Return from Tomorrow or My Life After Dying by Dr. George Ritchie are all examples of first hand NDE accounts that make interesting reading. The difference in their perspectives is important to note as Eadie is Mormon, Brinkley does not identify himself with any particular faith but is spiritual, and Ritchie was a Christian. NDEs cross religious boundaries–while many reports of NDEs are by Christians, Muslims, Hindus, and Jews have reported them as well. No one religion has a monopoly on NDE experiences.

It was Dr. George Ritchie's personal NDE account that inspired Dr. Raymond Moody to seriously investigate near death experiences. As a 20-year-old army private with pneumonia, Ritchie suffered cardiac arrest and was clinically dead for nine minutes according to records in the army hospital. The doctors recorded the time of death, and Ritchie's body was prepped for the morgue.

While his physical body lay in a small storage room, Ritchie claimed to have walked the hospital corridors until an orderly coming the opposite direction couldn't see him and walked right through his body. Ritchie said before returning to his physical body he travelled through a dark tunnel toward a bright light and there he met Christ and watched a review of his life. Years later, after he described his experiences to Dr. Moody and inspired Moody's book Life after Life, Ritchie wrote his own NDE account in My Life after Dying.

NDEs are not all the same. Eadie and Ritchie both described meeting Jesus in heaven. Dannion Brinkley thinks of the entities he met while dead as keepers of books of knowledge and seemed reluctant to refer to them as angels or God. But Brinkley's experience definitely changed his life as much as any other NDE recipient. Based on his own description Brinkley changed from a narcissistic person to a man who volunteers at hospice care. There he uses his special knowledge he acquired during his own NDE to help others nearing death make their transition from this life more easily.

First person NDE books make very interesting reading, but they may be more easily criticized for their lack of scientific data than Transformed by the Light. It may be argued personal NDE accounts are inherently biased. The author recounting his own NDE has an apparent conflict of interest because as the person claiming to have had the experience stands to profit from his story. Therefore, he or she might be tempted to embellish the details.

These books depict honest representations of what individuals honestly believed they experienced as a NDE. That is clear. As far as the scientific effort known as the Transformations Study was concerned, whether the patient actually saw God was irrelevant information. The study only cared about whether and how they had changed after their NDE. At the risk of sounding repetitive, participants in the Transformations Study had nothing to gain by lying or embellishing the truth because they weren't paid for their time.

That is a very important point because if they weren't lying, the only other choice is they were telling the truth. Documentation was required to validate their claims medically, and interviews were conducted on both the study participants and any available witnesses to confirm every detail before being included in the study.

There are far too many NDE accounts told by too many people from all corners of the world for skeptics to be able to ignore without further study, and too many NDEs have been told by very credible people to be summarily dismissed as fraud or fantasy. The possibility these NDE events are hallucinations somehow manufactured by the dying mind has been discredited by the investigators in both the Dutch study and the Transformation Study. Evidence supporting the positive impact of these experiences on the NDE patients was substantial, significant, and measurable.

In Transformed by the Light, Dr. Morse noted that people who have had a NDE tend to perform more volunteer work, have fewer physical ailments or general complaints, eat better, take fewer medications, give more to charity, and often take careers in helping professions such as nursing, or teaching special education. In short, they live happier, more fulfilled lives than the general population. Dr. Morse was careful not to claim when these people "died" they really went to heaven and met God, but he does state the NDE accounts "have been independently verified by witnesses whose testimony would be accepted in any court of law in the land." [59]

NDE COUNTER ARGUMENTS

Morse acknowledged that experiments during surgery performed by Canadian neurosurgeon Dr. Wilder Penfield also produced out of body states and the sensation of traveling through a dark tunnel by touching the right temporal lobe with electric prods. Users of psychedelic drugs such as LSD or peyote have reported out of body experiences as well. However, these experiences were induced by artificial stimuli such as an electric prod or a drug. Artificial stimulation failed to reproduce the sensation of being in the light, which appears unique to the NDE.

In addition to the life changing transformations that occur as a result of the NDE, Dr. Morse made note of some unusual changes in the electro-magnetic fields in a disproportionately large number of NDE study participants. Roughly twenty-five percent of NDE survivors said they can no longer wear a watch.

One participant purchased a couple of expensive watches that stopped working when he put them on his arm, but worked perfectly without repair when he gave them to his sons. Another man claimed that his watch worked fine as long as he left it in his dresser drawer, but as soon as he wore it the batteries died.

Dr. Morse's work should not be criticized for bias because he did not recount a personal experience. His Transformations study carefully documented how people were forever changed by nearly dying. The conclusions drawn by the Transformations study were supported by facts obtained using the scientific method.

Morse readily admits he now believes NDE stories are true because they "are told with such beauty and simplicity by children and adults alike that have nothing to gain from making them up." [60] Dr. Morse admitted he originally thought they were hallucinations caused by medication or lack of oxygen in part of a psychological defense for the fear of dying.

He changed his mind about life after death after stories told by his patients made him open his own mind. It took the straightforward, shockingly honest story of a seven-year-old little girl named Katie to rock the foundations of his knowledge and understanding. Morse openly acknowledges further research work needs to be done. He encourages other medical professionals to follow suit and study the near death experience more.

Dr. Morse expresses regret the medical community routinely dismisses and devalues NDE stories. When patients are treated as wacky and delegitimized rather than comforted and treated with respect, both patient and doctor are harmed. Some in the medical profession argue NDE stories cannot be real without giving a valid reason why not.

INCONCLUSIVE CONCLUSIONS

Morse sought and received help from the most conservative people he could find in his profession to negate the kook factor for his research. He employed the services of both medical experts who had done research on NDEs as well as some who had no exposure to the field whatsoever. Dr. Moody cannot be faulted for making his effort to conduct a fair scientific study of the NDE.

Richard Dawkins has said abiogenesis only had to happen once for the rest of evolutionary theory to be believable. Applying that same logic to the NDE, only one of these experiences has to be real to give us plenty of reason to believe in life after death, God, and Heaven. Death is the end of physical life but emerging evidence suggests our spirit and consciousness continue to exist afterward.

10 MISUNDERSTANDING GOD

Ah, God said to Abraham, "Kill me a son"
Abe say "Man, you must be putting me on"
God say "No." Abe say "What?"
God say, "You can do what you want, Abe but
The next time you see me comin' you'd better run"
Well Abe say, "Where you want this killin' done?"
God says "Out on Highway 61."

– Bob Dylan, Highway 61 Revisited

Why the hostility toward God of the Old Testament?

I can't fault atheists for being confused about some passages regarding the God of the Old Testament. I am also confused by things I read in the Bible from time to time. If I ever claimed I could thoroughly explain God to somebody, I'd be lying. Many passages and stories, just like the story of Abraham and Isaac mentioned in the song above, have dubious meaning. Of course, Bob Dylan took a little creative license with the story, but the basic premise is accurate–God did tell Abraham to offer his son Isaac as a human sacrifice.

It's very important to note that if we assume God exists, we must realize we are not worthy to judge Him. Questioning God is different; asking why so-and-so had to die so young or why such a tragic accident had to occur or why that storm hit where it did and hurt so many people is normal and natural. Expecting an answer is not. If you are a parent, do you feel obligated to explain to your children every decision you make? Wanting to ask someone a question and demanding an answer are two different things.

In the case of asking God to explain His motives, it will likely do you no good. God feels superior because He Is Superior and does not condescend to our demands that every detail of our existence be explained. Things came to exist because God Is Superior and not required to acquiesce and answer to the petty demands of His inferior creations.

On the one hand, the evolutionist must be able to present a coherent and comprehensive alternative to God's creation in order to be considered credible by the mass public. Although the systematic assault on faith by pseudo-scientists has increased the number of atheists and agnostics into double digits, the overwhelming majority of people still believe in some sort of God because the evolutionist's argument ultimately requires an even greater leap of faith.

On the other hand, creation gives a comprehensive explanation for the origin of the universe, the origin of life, the introduction of different species of life and the sapient human. Evolutionists browbeat about the minute details lacking or apparently confusing to some in the Judeo/Christian creation account while ignoring the huge leaps of faith in their own religion they call science. Science isn't bad–it's given a bad name by pseudo-scientists just as religion has been given a bad name by events like the Spanish Inquisition.

From "What is Evolution?" through "The Crux of the Debate" we discussed evolutionary theory and where creation science differs. It was clear where I stood on the points of evolutionary theory that are valid (Big Bang and natural selection) and logical and which points are not (speciation and abiogenesis). In "Personal Experience" I offered sincere testimony that I believe I was visited by God in my room one night, and that I also believe I saw evidence of paranormal activity on multiple occasions when visiting one particular friend's house.

In "Spooky Stuff," I wrote about documented cases of psychics assisting police efforts to solve crimes and help people. They demonstrate solid evidence to accept the fact unknowns do exist an inexplicable phenomena really do occur. "Reason to Believe" included accounts of the paranormal experiences known as NDEs during which people die and come back to life and gives reasons to believe in life after death and that heaven and God are real.

The consistency and authentic recounting by former patients of their experiences and more importantly, how their lives changed as a result were documented by scientific research. The stories themselves cannot be proved via scientific method, but evidence to support belief in a supernatural God is overwhelming when viewed in cumulative form.

If these stories are true, and we have no reason to believe otherwise, isn't that at least circumstantial evidence of God on par and equally as strong as any fossil or astronomic evidence of the Big Bang?

I have explained why I share evolutionist's skepticism about the true elapsed period of time for creation and my belief that time was not measured in Earth days but in God days. That is insufficient reason for rejecting God as Creator.

EXPLAINING THE INEXPLICABLE.

Coincidentally, if one happens to mention God has supernatural qualities, Richard Dawkins will invariably attempt what may be perceived on the surface to be a clever argument but when closely examined proves to be a dishonest one.

In that momentous interview with Dawkins, when Colbert pointed out the obvious, "God is outside of time." Richard responded, "That's so easy. You get away with that and then you can explain anything. I can explain it by saying you get to complex things like you by slow, gradual degrees and that's the only ultimate explanation that will work. You can't just suddenly "magic complex" things like God into existence." [61]

To his credit, Colbert tried to make sense of what Dawkins just said, asking, "But let's say your book is intelligently designed."

Dawkins interjected, "It is, by the way" for a quick laugh by the audience.

Colbert continued, "But the universe is not intelligently designed, so you're saying the universe naturally came into existence and continues existence by the natural laws of thermodynamics and physics, the laws of gravity and energy produced you and through you produced this book that proves it has no natural Intelligent Design?"

Colbert appeared genuinely confused asking the question.

Dawkins stonewalled his answer, beginning by saying, "Let's take this step by step."

This time Colbert interrupted: "I don't think we have time for step by step. You could just surrender, and we could go."

Dawkins was undeterred: "You were right as far as when you got on to life. Life is a very special thing. Life starts naturally then increases in complexity by slow, gradual degrees–that's Darwinian natural selection, and ultimately that's the only explanation that will work."

"That's because God breathed into it," Colbert counter-argued.

Dawkins was pleased to quote himself and said, "Oh, no, that is at best a superfluous hypothesis, and at worst a highly unparsimonious one." [62]

Dawkins cloaked abiogenesis and speciation in disguise as Darwinian natural selection in a slick but completely false statement. The very thing evolutionists accuse skeptical people of doing (which they do but accidentally) Richard Dawkins does deliberately! He referred specifically to abiogenesis and speciation both and called them natural selection, giving credibility to two preposterous and unproven theories by calling them the one "proven" theory of evolution.

His command of the English language was rather impressive but his argument was completely false. The thing is Dawkins knows better. He knows the only way he can sound convincing is to misrepresent two component theories of evolution as a third. The interview accentuated the atheist's absolute refusal to consider the only other alternative, which happens to also be the only completely viable one. When God accounts for the Big Bang and offers the only truly plausible explanation for abiogenesis and speciation, natural selection does make perfect sense. When God is omitted, the theory stands alone with nothing to explain a large majority of what needs explaining.

Distressingly, there is a movie with the ironic title Creation starring Paul Bettany and Jennifer Connelly as Charles Darwin and his wife Emma, which has yet to get a distributor in the United States. [63] The producer claims the film can't be distributed due to creationist bias toward belief in a strict six Earth day period of creation and a backlash from conservative evangelical Christians who staunchly believe in strict creationist theory, which is an interesting charge considering the recent court ruling that refused creationist theory into school under any circumstances.

If true and not merely a publicity stunt to gain attention for the film (which worked, because I read the article), it is a sad commentary on the reality that strict creationists are equally close minded in their beliefs as evolutionists and, therefore, I fear neither side of entrenched theorists will receive Divine Evolution with an open mind.

I've provided strong arguments for consideration of this hybrid theory. The evolutionists complain that creationists seek a "God of the Gaps," but they have black holes in their arguments into which all logic disappears. The blind bias of the evolutionist against any consideration of God's existence makes it practically impossible to have this conversation with them about evolution theory. But God is larger than the topic of evolution or creation. It is a fundamental misunderstanding by the evolutionist of who is the "I AM," the Alpha and Omega.

However, atheists will have still more objections to God. The concept that God is omniscient, omnipresent, and omnipotent is one troublesome concept about which Dawkins scoffs. The triune God is another theory that is difficult to conceive and, therefore, difficult to believe for Dawkins, who expresses his incredulity.

THE NATURE OF GOD

God is frequently described as omniscient (all knowing), omnipresent (everywhere) and omnipotent (all powerful) in Christian teachings. But those words are not in the Bible. What does it mean to be all knowing or everywhere?

The triune God means God the Father (Yahweh), God the Son (Jesus the Christ or Messiah) and God the Holy Spirit. Remember Jung added a fourth dimension called the God Within and called the relationship the quaternary. How can God be three distinct entities and one simultaneously?

Because this concept is one I least understand, I will first deal briefly with the concept of the **Triune God**, meaning God the Father (Yahweh), God the Son (Jesus the Christ) and God the Holy Spirit are separate entities but all part of the same God. That's a difficult idea for me to comprehend. I know Jesus said, "I am in the Father and the Father is in me," but he also said only the Father knows the day and the hour of His Second Coming. I've read interesting theories about how we each are sparks from the Eternal Flame of Life we call our Creator God, but rather than spend a great deal of time explaining someone else's theory, I will defer and admit I can't explain the Triune God very well.

Now you know brief definitions for each term, but what do the words really mean? Does omnipresence mean God is literally everywhere at every moment of the day? When the Bible says God hears our prayers, does that mean He can hear every person who is praying at the same time? Let's look at each of these three concepts about God individually.

IS GOD OMNIPOTENT?

The self-described liberal Tony Campolo has said God is not and His power is overrated. Writing on Belief.net, Campolo started out okay as he rejected the God thought to be punishing the people of New Orleans by inflicting Hurricane Katrina upon them.

Campolo wrote, "When the floods swept into the Gulf Coast, God was the first one who wept. Perhaps we would do well to listen to the likes of Rabbi Harold Kushner, who contends that God is not really as powerful as we have claimed. Nowhere in the Hebrew Scriptures does it say that God is omnipotent." [64]

He reiterated my point that the Old Testament does not use the word omnipotent, but it clearly says God is all powerful through the accounts of His deeds. The creation account alone is ample evidence of God's immeasurable power. Genesis 1:3 reads "And God said, "Let there be light," and there was light.

All God had to do was speak and light (matter or energy) was created. If He can do that, God is indeed all powerful, and His might exceeds our wildest imagination. The Bible says God stretches the heavens. Moses didn't part the Red Sea, he used power granted by God to cause the waters to part. God owns and controls the physical universe. There are ample Bible verses wh

IS GOD OMNISCIENT?

There are Bible verses which indicate God is indeed all knowing, but that doesn't necessarily mean God knows everything in what computer programmers call "real time."

Real time means now, immediately. Processing time is how long it takes for a computer task to occur, where real time means how much CPU time elapsed. Several seconds of processing time may elapse while a program runs, but only a fraction of a second of real CPU time might be dedicated to the task.

Despite protests of Richard Dawkins and his friends, there is every reason to believe God is truly a Being unconstrained by time. Time did not exist before the Big Bang, but God did. God has no beginning and no end, so to Him everything is in "real time."

The evolutionist doubts God's omniscience because God doesn't appear to answer every prayer. In fact, some evolutionists have gone so far as to perform stunts like conducting the Great Prayer Experiment, where a scientific study was allegedly conducted to determine the effects of prayer on the recovery of heart bypass surgery patients.

Dawkins mocks the results of the study for showing no improvement of patients receiving "special prayers," but I must take exception to the premise of the study itself. After his final temptation when Satan challenged Jesus to throw himself from the temple roof because God would not allow harm to befall the Messiah, Jesus answered, "It is written, you should not put the Lord thy God to the test." A scientific experiment to check for the effects of prayer is an attempt to test God. The Great Prayer Experiment was a poorly conceived idea from the start.

The Bible says our Creator knows the number of hairs on our heads, so He definitely knows everything about His creation. He knows the number of feathers on the sparrow and cares about their welfare. God knows everything we do, but I think it's possible that is because we tell God after we die. Remember the life review from the NDE accounts? This argument is intertwined with the theory of God's omnipresence.

Our memory or subconscious has recorded everything we've ever done. Like Edgar Cayce, we may not remember everything we know, but we still know it. It's in there somewhere. Our data retrieval mechanism from the conscious mind may not operate at peak efficiency when we're alive, but after we die those memories are available from our subconscious for our life review, when we tell God every gory detail of every deed we performed on earth, both good and bad. Of course, this theory is unnecessary if it also proves that God is omnipresent, but this does provide an alternate theory to explain how God could judge our every deed and not be everywhere at all times.

IS GOD OMNIPRESENT?

Is God everywhere at all times? It makes perfect sense that God may be anywhere at any given time, but is He everywhere at all times? Remember, when we think of "everywhere" we think of everywhere on earth, but God is not confined to the earth so everywhere would mean everywhere.

Again, pardon my opinion–as my dad used to say (with one different word) an opinion is like a derriere; everybody has one. But I do think my opinion makes a good bit of common sense, so I dare put it forth. I do not think God is everywhere at all times, nor do I believe the Bible says God is. In fact, the Bible suggests the opposite is true.

In the book of Genesis, Yahweh told Jacob that He would remain with him while he journeyed from Bethel to Paddam Aram and back and would not leave Jacob during that period of time. The implication of God to make such a promise is there are times when He isn't always in such close proximity. [65] In other words, if God is everywhere simultaneously at all times, why would God have to say to Jacob what He said?

God anointed Moses to lead the people of Israel from Egypt to the Promised Land. But after sending Moses to Egypt, according to the Bible, God met Moses at Midian with the intention to kill him. Moses had neglected circumcising his son after being chosen by God. Yahweh chose Moses to lead His people from Egypt to Canaan but was obviously ready to change the plan if Zipporah had not intervened and performed the circumcision. She smeared the blood on her husband's feet and thus God spared Moses' life.

God wasn't merely being a stickler for detail. He chose Moses to lead the people of Israel and God required full trust and obedience as he handled the responsibility for a nation of people. Moses knew it was his duty to have his sons circumcised according to Hebrew tradition. His reluctance to honor the traditions of his faith indicated he lacked complete trust in God to fulfill His promises, knowing full well the awesome power of the Egyptian pharaoh by mortal standards. Moses demonstrated understandable reluctance to face his worst fear. He fled Egypt in fear of his life and at first had difficulty accepting his role and responsibility. Disobedience to God and his failure to follow strict instructions cost Moses his entry into the Promised Land, but that was an earthly punishment. Moses surely earned his place in heaven by following God's direction for his life more often than not.

By not circumcising his son, Moses hedged his bets. His children would assimilate into a foreign culture more easily if they weren't circumcised and if God didn't follow through on His promises to Moses every base would still be covered. Moses demonstrated a lack of faith and trust in God.

Moses was predestined to lead Israel only in the sense God chose him. Had Zipporah not saved the day, we'd be talking about Aaron or Dudley leading the people of Israel to the Promised Land. Moses might well have become a biblical footnote. God was not fooled by Zipporah's action; He has the ability to discern and extract truth from our consciousness. He recognized Moses had the support of a strong wife who would not allow him to stray in faith very far. But he did continue to stray.

There is a second moral to the story of Moses. He was flawed in spite of being a great leader of Israel and a mortal man through whom God performed spectacular miracles. Yet he was denied entry into the Promised Land because he and Aaron disobeyed God and broke faith at the waters of Meribah Kadesh in the Desert of Zin in the presence of the Israelites. Moses led the Israelites to the Promised Land, but the closest he came to entering was from Mount Nebo where he saw into Canaan and he died. He was buried in Moab; not even his body made it to Israel.

David was anointed by God as King of Israel, defeated Goliath and was one of the great men in the Bible–until he committed adultery and murdered the husband of his mistress by proxy. It cost David his existing family and great personal loss. David was a warrior king, the poet who wrote most of the book of Psalms and a great leader for the people of Israel, but he grew proud and allowed that pride to lead him astray. He sinned greatly against God, but it took the prophet Nathan's visit for David to open his eyes. He spoke in a parable about a rich man stealing from a poor one, and when David's anger burned against the injustice Nathan revealed David was the guilty man for stealing Uriah's wife, Bathsheba, a crime he compounded by making sure Uriah was killed in battle. Nathan predicted the consequence of David's sin was the child born from his adultery would die.

Each story is very different, but both ultimately make the same point. **Actions have consequences.**

I would anticipate Dawkins' objection to the story was that David was not killed but his son was. In fact, the child was not the only son David lost, because his adult sons rebelled against him. One son kills another for raping their sister and then is killed himself by Joab, one of the commanders in David's army. [66]

COSMIC CHECKMATE

Belief in God does not require fully understanding of God. For example, consider the story of Job. The story reads as if Job, his family and possessions were practically pawns in a cosmic chess game between God and Satan. Job was put to the test, yet he endured with patience and faith and serves as a model for us to follow.

At his darkest hour, Job still could not be enticed to curse God; instead he pleaded for mercy and a quick death. Job didn't doubt or question God's authority, but surely he wondered what he had done to deserve the punishment he endured. The answer was, not anything. His story tells us we should not blame God for our misfortune nor assume we've done something wrong, if we suffer harm.

More importantly, we should not assume God doesn't exist if we aren't immediately punished for doing wrong, either. Apparently, He has means of dealing with aberrations in society like John Wayne Gacy by working through people such as Carol Broman and Dorothy Allison as unusual instruments for good.

I must confess I have often wondered why God allowed Satan to torment Job. I also wonder, if God literally "wrote" the Bible as some theologians claim, why are there so many stories included which are unflattering? As I pointed out in The Crux of the Debate, change the meaning of a single word in the creation account and the period of time changes the time of God's creation from six rigid days of 24 hours to an unknowable period of time.

It is rather obvious that the sophistication of biblical authors improves significantly over time. Because God is credited as author but humans actually did the writing, one can reasonably assume the change in sophistication is due to improvements in the education of the human scribe while God remained constant. It seems reasonable to conclude that God communicated to each writer in language and terms both he and his contemporary audience would understand, but also in a form that communicates God's will and message for future generations.

The Bible is such a unique book because it has so many meanings and yet is terse and informative at the same time. There are several instances of multiple accounts of a single event in the Bible given from two or more perspectives. For example, in 2 Kings the author spends all of chapter 20 describing Hezekiah's illness and God's miracle cure, confirming His grace by moving the sun backward in the sky 15 degrees as an indisputable sign to Hezekiah of God's power to fulfill His promises. The Kings account emphasizes God's power, might and grace and paints a less flattering portrait of Hezekiah.

By contrast, the author of Chronicles covers the reign of the same king Hezekiah from a completely different perspective. It is more flattering to the king as it showed good works to please God as the motivation for Hezekiah's accomplishments. The account in Chronicles glorifies the works Hezekiah performed on behalf of the people of Israel such as channeling fresh water into Jerusalem and leading the people closer to God, but both accounts agree that pride led to Hezekiah's downfall and diminished his legacy. When he exposed his gifts from God and the riches in his treasury to the Babylonians, they became motivated to steal them. The Bible says "God left him to test him and to know everything that was in his heart" [67], implying yet again that God is not literally everywhere all the time.

When I contemplate the meaning of Job, it still bothers me to think about the 'cosmic chess game' apparently played with Job's family and prosperity as the sacrificial pawns in the game—what I can't get over is the obvious cooperation between God and Satan. For whatever reason, God gives dominion over Job to Satan to be put to the test. The reason for Job's suffering is impossible to comprehend. It's as if Satan suckered God into agreeing to Job's torture.

But that's the story of Job on one level. Job is a story of faithfulness by a man in spite of challenges and afflictions. Job and Hezekiah are both tested by God to find what is in their hearts. God's perspective as a timeless and eternal Being is not the same as ours, and He knows what is or can be in store for Job both for the remainder of his earthly life as well as what comes afterward.

Job's story does have a happy ending, at least for him if not his dead children, as his wealth is restored and he has another family. Even his friends who come to commiserate with Job instead accuse him as responsible for the travesties that befell him. These "friends" soon found themselves needing his intervention in prayer to escape God's wrath for their criticisms of Job.

Job didn't "deserve" his afflictions, nor did his children and livestock "deserve" to die from the perspective of human judgment. On the other hand, theoretically what do we know about anything? Everything that lives will eventually die, or at least the physical body they inhabit will cease to function. Death is painful to the living in part because we don't understand it and therefore we tend to fear it.

It may be true that death and the opportunity to meet God are far superior to what this life can offer, but we have to die to find out for sure. Waiting to find out naturally seems the prudent decision. Relishing each moment in this world for being the precious gift that it is seems perfectly reasonable and appropriate. It is mildly interesting to note that after his ordeal Job had more wealth than before, and the same numbers of male and female children were born to replace the children Job lost, not that it is really possible to replace one child with another. That sucked for Job's first round of progeny, but that is probably why some people believe in reincarnation in order to have hope in this world for those who die prematurely. God must have comforted Job's loss of his first family by some knowledge not revealed in his book.

If I had to summarize the book of Job in a single sentence, I'd have to say that just because the story is difficult to understand doesn't mean it isn't true. When told by his wife that it was time to curse God and die, he called her a fool and asked "Shall we accept good from God and not trouble?"

The story of Abraham (the real one, not Bob Dylan's version told in the introductory song lyrics at the beginning of the chapter) really bothers Richard Dawkins. Here's a shocker; the story bothers me too. I don't understand God's motives as the story is explained either except that God wanted to test Abraham's faithfulness.

God told Abraham to take his only son Isaac to the land of Moriah and sacrifice him as a burnt offering. He stopped Abraham before Isaac was slain when it was obvious that Abraham would not hesitate to obey. When God told Abraham what he wanted, the Bible says Abraham got up early the next morning and saddled his donkey. Every step of the way surely was agony for Abraham, but up until the moment God stopped him, Abraham was willing to faithfully carry out God's commands.

Richard Dawkins surely takes umbrage with the fact God admitted He tested Abraham to make sure he feared Him. The sad truth for humanity is fear breeds respect and obedience, and comfort spawns a lackluster effort and a lackadaisical attitude toward authority. God gives us what we need, not what we want. It is true that you can't always get what you want.[68]

Dawkins also refers to the story of Jephthah[69], pointing out the apparent contradiction that Isaac was spared but Jephthah's daughter was not. Jephthah was an exiled warrior chosen by the people of Israel to lead their defense against the invading Amorites.

Jephthah had been driven from Israel because his mother was a prostitute, and he had no inheritance there. The people of Israel had been disobedient to God and were punished for worshiping false gods. When the Amorites attacked Israel, the people repented and reached out to Jephthah, promising he would rule them after defeating their enemy. In exchange for their promise, Jephthah made an oath to God that he would sacrifice as a burnt offering whatever came from his house to greet him upon his victorious return home. It was not an oath Jephthah would give lightly, but he came to regret it just the same.

Jephthah was afraid to renege on his pledge to God. Why God is considered by Dawkins to blame for the death of Jephthah's daughter and not Jephthah himself? As unpleasant as it sounds, the distinction must also be noted between the stories that Abraham was tested when God demanded the sacrifice of his son but Jephthah made an open promise that culminated in the sacrifice of his daughter. And no, God isn't sexist.

We have also established the theory of God's omnipresence does not stipulate His presence everywhere at all times, but His ability to be anywhere at any time. It is possible God wasn't there at the moment of the sacrifice.

There is also the matter of personal responsibility. The implication of Dawkins' argument is if God did exist, He would never allow anyone to die for any reason. If that were true, not only would we lack free will, we would never learn if we fail to suffer consequences for our actions. Jephthah made an oath he later came to regret, but God was faithful to the people of Israel and he was compelled to be faithful to God.

These Bible stories make the reader wonder about God, creating doubt Dawkins attempts to exploit with enthusiasm. These stories require some deep thought for any type of understanding–Jephthah saved the people of Israel from the invading Ammonites, and his reward was that he had to sacrifice his only child to God?

THE LIKELIHOOD OF IMPROBABILITY

There's an old programmer's joke about using comments (non-executable statements that serve to explain the purpose of the program) into your source code that goes like this: don't bother with comments, because if the code was hard to write, it should be hard to understand.

Richard Dawkins acknowledges the Bible is a great work of literature. He argues it is not a history book and therefore has no historical truth. Dawkins mentions a number of passages from the Bible that he cites as great literature to be treasured, just not believed. He certainly wants no part of biblical morality. Dawkins suggested Abraham was mythical. He expressed doubts Jesus was a real person, but stipulated even if he was not a fictional character, Dawkins was certain he was not God incarnate.

But Dawkins exposes his true bias toward God in one quote most notable for the degree of hostility in the sacrilegious rant he unleashed to begin a chapter in The God Delusion titled The God Hypothesis. He begins an insult of the Judeo/Christian God with possibly the most offensive tirade against Yahweh ever captured in print: "The God of the Old Testament is arguably the most unpleasant character in all fiction: jealous and proud of it; a petty, unjust, unforgiving control-freak; a vindictive, bloodthirsty ethnic cleanser; a misogynistic, homophobic, racist, infanticidal, genocidal, filicidal, pestilential, megalomaniacal, sadomasochistic, capriciously malevolent bully." [70]

Most of the attributes Dawkins just assigned to God are actually descriptions of human behavior apparently done in God's name. There are a couple of references in the Bible to God's instructions to wipe out a certain race of people, but some experts have claimed this was due to common practices by these foreign nations of child sacrifices to a pagan god. This practice angered the one true God and inspired the order.

Dawkins' God Hypothesis says the thought that "there exists a superhuman, supernatural intelligence who deliberately designed and created the universe and everything in it, including us" [71] [Emphasis original] is preposterous. I have proposed Divine Evolution to support his God Hypothesis. Whether or not my argument is convincing depends on which position you, the reader, find most compelling.

Remember, Dawkins claims the supernatural does not exist, but I have given testimony to the contrary using personal experiences as the basis for my belief in paranormal activity. I've seen too many different things happen a number of occasions that I cannot logically explain. I expect a professional illusionist like David Blaine to be able to convince me I am seeing things that aren't really happening, but even those examples have a visible performing artist and the same cannot be said for the things I experienced over a period of several years when visiting Jim Nettles' house in Savannah.

To support his own theory and help advertise The God Delusion, on October 23, 2006 Dawkins posted on Huffington Post an article titled Why There Almost Certainly Is No God in which he wrote: "We explain our existence by a combination of the anthropic principle and Darwin's principle of natural selection. That combination provides a complete and deeply satisfying explanation for everything that we see and know. Not only is the god hypothesis unnecessary. It is spectacularly unparsimonious. Not only do we need no God to explain the universe and life. God stands out in the universe as the most glaring of all superfluous sore thumbs. We cannot, of course, disprove God, just as we can't disprove Thor, fairies, leprechauns and the Flying Spaghetti Monster. But, like those other fantasies that we can't disprove, we can say that God is very very improbable."

I don't really fault the atheist for not understanding the true nature of God. Who does? But I can fault them for not even trying to figure out that God is real. I fault those who mock, insult and ridicule those of us who believe in a divine power. Dawkins not only had the audacity to put his insult of Yahweh into print, he hawks T-shirts on his web site trying to profit from his foul aspersions on God. Dawkins amazingly complains about hate mail he receives from Christians as if he's done nothing provocative. He pretends his book was written to discourage the pervasive Christian from spreading his dogma but helps sponsor buses advertising the freedom and benefits of atheism. In short, Dawkins wants the freedom to espouse his personal views while wishing to restrict or silence his philosophical counterpart.

The real objection these people seem to have is a reluctance to have their behavior judged. Who wouldn't want carte blanche to do as you please with no fear of repercussion? Even if you don't want to harm others in the process, surely having every whim indulged has appeal? The problem is, inside we know the difference between right and wrong. We all know adultery is wrong, but some of us let their sexual desire overshadow their judgment and succumb to self-indulgent pleasures. They justify their actions with "everybody else does it" or other rationalizations, but the truth is they know it is wrong just like everybody else, but are too weak or don't care.

And before it sounds like I'm being judgmental about the adulterer, I'm just giving an example, not picking on any particular sin. My sins aren't any better than theirs. My desire to remain married to my wife exceeds my desire to perform mating rituals with strange females with a short-term strategic objective in mind. It isn't worth it. But because it is a sin doesn't mean that's why I don't do it, because lying is also a sin, and I do that more often than I care to admit. Even as I write this, it is tempting to tell another lie about how often I lie.

I'll be happy to explain my beliefs and perspective and listen to the atheist point of view as long as the discussion remains cordial. Insulting my God in smug, condescending and inflammatory rhetoric is not cordial debate. Once the debate degenerates into childish insults, the one using them has lost the argument because he has run out of ideas.

You can almost hear glee in Dawkins' writing as he made this assault on God's character. What's really funny is just how much of a hypocrite Dawkins was about the subject of T-shirts with inflammatory slogans.

Because Dawkins sells T-shirts on his website with his lengthy insult of Yahweh quote on the front, one would have to assume that he doesn't think the comment is the least bit offensive or objectionable, otherwise, why would he put it on a T-shirt? In The God Delusion Dawkins told the story of a twelve-year-old kid named James Dixon who won a court case in Ohio by claiming freedom of religion.

Winning his case gave James the right to wear his t-shirt to school proclaiming, "Homosexuality is a sin, Islam is a lie, abortion is murder. Some issues are just black and white!" The school had banned the shirt, James's parents sued, and the Supreme Court ruled the shirt permissible under the First Amendment. Dawkins asserted the shirt was tantamount to hate speech, and claimed it was unprotected in the Bill of Rights. Dawkins did not mention whether he considered his own t-shirts to be hate speech, and I suspect he'd have been pleased if it was a shirt with his infamous God Hypothesis insult.

Dawkins doesn't consider offending, mocking and ridiculing people of faith as hateful, but a twelve-year-old kid wearing a t-shirt with a similarly provocative expression is evil incarnate, apparently. Personally, I wouldn't buy a shirt like James's for my son, but I wouldn't buy him Dawkins' shirt, either.

Dawkins concludes his God Hypothesis by saying not only is belief in God a delusion, but a pernicious delusion.

The vehemence with which Dawkins expresses his opinion of God's likelihood of existence draws to attention the degree of bias that impairs his judgment and jades his impartiality. Actually, Dawkins doesn't bother to exhibit an interest in showing impartiality whatsoever.

Painting all skeptics of his works as strict creationists allows Dawkins to refuse to debate them, thus avoiding most legitimate criticism of his work. Dawkins states that he named his book as such because delusion is defined as a fixed, false belief, specifically in any 'supernatural' gods.

But what other kind of God is there? I surmise he means the god of 'Einsteinian' religion as he coins it when he wrote about "the unbounded admiration for the structure of the world so far as our science can reveal it."

Dawkins goes on to quote Einstein and other prominent scientists he asserts to be atheists, including Carl Sagan, Stephen Hawking, and others who ostensibly are supposed to impress the reader. We have previously established that smart people sometimes say stupid things.

When Dawkins 'fleshes out' his God Hypothesis, he objects to both the premise that the Abrahamic God "…not only created the universe, he is a personal God dwelling in it, or perhaps outside it (whatever that might mean), possessing the unpleasantly human qualities to which I have alluded."

It is unfortunate that his hostility will probably blind Dawkins to the possibilities offered by this hybrid theory. If he ever did consider the possibility, he'd also have to consider how to go about offering a sincere apology to God for the insults he spewed.

11 FICTION SCIENCE

Richard Dawkins has said creation and evolution are not equi-probable theories, and he's right. Only two competing theories attempt to answer the question of how life came to be. While I agree with Dawkins these theories are not equally probable, I disagree with his assertion that evolution is the only theory that makes any sense. On the contrary, creation is **much easier** to believe because evolution doesn't really mean anything.

No synergy is created by combining the theories of the origination of the universe, the origination of life, the spread of life, and the development of life today together without God. Attempts to bundle these into one cohesive theory in order to compete with the theory of creation requires the suspension of all doubt and that we overlook the weak links in evolutionist theory.

The smug evolutionist may pounce on the unsuspecting skeptic who questions one of the four component theories and calls it evolution, but the semantic error fails to validate evolutionist theory. It serves as opportunity to humiliate innocent "less enlightened" people, but it does not advance evolutionary theory because the reality is that evolutionary theory is a convenient term_packaged for mass marketing.

I've already stipulated the Big Bang Theory and Darwin's theory of natural selection are plausible. However, a few carefully selected facts fail to provide a comprehensive alternate theory to creationism. Evolution may gain acceptance as fact from sheer repetition by talking heads and pseudo-scientists dutifully repeating the prevailing opinion, but that doesn't constitute actual fact. Every scientific theory regarding the origination or perpetuation of life on earth ultimately fails absent a Creator.

The Big Bang Theory fails without a Creator providing the matter. Abiogenesis fails without the breath of life from God. Speciation fails absent

Intelligent Design. Natural selection appears solid until it alone is compared to creation as a comprehensive theory, then it is exposed as only a definition and not an explanation. Absent the other three prongs of the theory, natural selection simply becomes another way of describing gene mutation commonly observed as heredity. It is common ground that can be equally claimed by evolutionist and creationist beliefs.

SNOBBERY OF NONBELIEF

Atheists believe Christians, Jews, Muslims, Hindus, and other religious people with strong faith are intellectually lazy. They believe religious people have taken the easy way out. By simply accepting, as reality, something they will never be able to see, hear or touch while inhabiting their mortal bodies, believers are told they are delusional and are to be pitied when not mocked, scorned and ridiculed.

Yet the atheists have no problem taking the easy way out when it comes to evolution. They choose to accept abiogenesis and speciation on blind faith even though evidence for those theories is an illusion at best. Just as Dawkins contends that living organisms only appear to be intelligently designed, I contend the so-called transitional fossils only appear to be evidence of speciation. Successfully bonding together a few amino acids does not measure up to the origin of life, but Berra and Dawkins want us to accept abiogenesis as fact. Remember the old adage "Appearances can be deceiving."

The problem for atheists is that they must be dishonest and cherry pick which fantasies they choose to believe to bridge the gaps in evolutionary theory, while scolding the religious for their faithful beliefs. Those are lies of omission. If we only claim the evidence we wish to believe is true and discard everything else as if it doesn't exist, we apply filters on the information we receive from our world capriciously. Failing to include fields of study like paranormal research or NDE research and accepting evolutionary theory as established fact shortchanges scientific progress.

Until Science can answer the questions about the Big Bang, abiogenesis and speciation without divine intervention, these evolutionist beliefs remain theories requiring as much faith, if not more than, the creationist alternative. To remain an atheist in light of the evidence provided in this book requires a willing suspension of disbelief of the natural observations we can make of the world around us as well as all the evidence of the supernatural world we cannot see.

Why shouldn't we give equal weight to the assertion that abiogenesis or speciation requires an Intelligent Designer? For one thing, advocates of the competing opinion do not want equal consideration and open-minded discussion of concepts they find threatening. Hostile evolutionists demand no weight or consideration be given to the view God's creation initiated the Big Bang and without God it would not have happened. If we try to see both sides, we see holes in purely scientific arguments, which require leaps of faith even exceeding creation science.

The public tends to trust science to tell us the truth. Put a lab coat and a stethoscope on someone and we tend to trust them more than we normally would trust a stranger. Society respects the hard work required to earn that lab coat, whether a true scientist is wearing it or an actor who plays one on TV. Looking the part and fulfilling the role are two different things.

Scientists stick to their guns and defend bad theory long after reason and common sense dictate other actions are more rational for two reasons: peer pressure and funding. It's tough to go against the flow of conventional wisdom and tell the world that global warming is a farce.

Bjorn Lomborg was an established and respected liberal scientist who believed completely in global warming and for that reason set out to disprove the theories of American economist Julian Simon. Simon claimed the world's environment was improving and not getting worse.

Lomborg actually studied the data and concluded Simon had been right for the most part. He dared publish his research in his book *The Skeptical Environmentalist*, and overnight turned from a rising star of the scientific community to an intellectual leper shunned by his peers. Michael Crichton described his treatment by the scientific community as "relentless ad-hominem attacks, which can only mean his conclusions were unobjectionable in any serious scientific way."

Crichton himself would come to understand how Lomborg felt better than anyone when his novel *State of Fear* was published. Critics in the scientific community howled. They complained the novel was fictional but said curiously little to refute the 31 pages of bibliographical sources of information Crichton provided at the end of his book.

The fact Crichton's well established writing style had been fiction based on in depth research into cutting edge medical and scientific technology for his entire career was apparently lost on those critics. The same brilliant analytic mind and captivating writing style, which helped Crichton become one of the most popular writers in recent memory suddenly appeared to be lacking sound judgment and a writer who injected flawed syllogisms into his work. The darling of the scientific community when he popularized science with novels such as *Jurassic Park* and *The Andromeda Strain* was excoriated for writing *State of Fear.*

The louder science protests debate of evolutionary theory versus creation, the more suspect their evidence and motives become. In a court case in the State of Georgia in 2005, the Cobb County school system, in suburban Atlanta, voted to put stickers on the cover of science textbooks that explained evolution was only a theory but was sued by the parents of a few students and the ACLU. The actual text of the sticker read: "This textbook contains material on evolution. Evolution is a theory, not a fact, regarding the origin of living things. This material should be approached with an open mind, studied carefully, and critically considered."[72]

Sounds reasonable, doesn't it? A federal judge didn't think so, calling the stickers an unconstitutional endorsement of religion. "By denigrating evolution, the school board appears to be endorsing the well-known prevailing alternative theory, creationism or variations thereof, even though the sticker does not specifically reference any alternative theories," U.S. District Judge Clarence Cooper said.

It is amazing that someone with so much power could be so ill informed. Judge Cooper doesn't seem to consider the documents on which he based his rulings. The Constitution guarantees freedom **of** religion, not freedom **from** religion. The mythical separation of church and state came from a letter by Thomas Jefferson to the Danbury Baptists, not the Declaration of Independence, Constitution, Bill of Rights or any other legislation passed by Congress into law.

Besides, as Judge Cooper pointed out there is no mention of alternate theory on the sticker, only the recommendation that students think about what they are taught. The sticker does clearly say that evolution is not a fact but is a theory, which is technically correct. Actually, we've seen evolution isn't really even a theory, it's a lump of theories cobbled together. But that is beside the point. Apparently the point is, we don't want our students to think for themselves anymore.

POWER OF PSEUDO-SCIENCE

Pseudo-scientists know so little about so many things, but protest loudly to the contrary. As William Shakespeare wrote "Methinks he does protest too much." I guess the theory behind the pseudo-scientist's rationale is, if you argue with enough conviction, you just might win friends and influence people. Or, as my dad used to say, if you can't dazzle them with brilliance, baffle them with B.S.

Pseudo-scientists rely heavily on computer models to form their theories. These models form the basis for their educated guesses. Computer models are fabulous for doing certain simulations and providing forecasts of future events as long *as the limitations of the model are understood and factored into the simulation program, and the results are properly weighted.* I am quite familiar with computer models and was considered to have achieved expert status as a developer after programming them for twenty years.

Based on knowledge derived from that experience, I wouldn't trust most computer models any farther than I could throw the developer(s) that wrote them. Simulation models are great for five-day weather forecasts. Technology and satellite observations have refined the data fed into the five-day forecast model, so it now predicts weather with reasonable accuracy, but only within that narrow range. Long-term forecasts are much more volatile, to the point where predicting next year's weather would be a laughable exercise.

A computer model might be able to forecast an event like an earthquake with sufficient and accurate seismic data, but global warming weather models are much less reliable and the results are just random guesses. So many variables have to be factored that the models simply won't work. Wind patterns, tides, and a host of factors determine our weather from day to day. Simulations might produce educated guesses, but they are still guesses at the end of the day.

I've always enjoyed reading science fiction, Jerry Pournelle, Ray Bradbury and Larry Niven are all great writers who immediately come to mind when I think of the genre. Isaac Asimov's *Foundations* trilogy of novels was the inspiration for Douglas Adams' *The Hitchhiker's Guide to the Galaxy.*

Some of the great Hollywood blockbusters had science fiction themes. The *Star Wars* movies by George Lucas were the highest grossing set of films of all time. The *Terminator* movies by James Cameron, Stanley Kubrick's *2001: A Space Odyssey*, the *Alien* films by Ridley Scott, and *Close Encounters of the Third Kind* by Steven Spielberg were all great movies and popular hits that captured the imagination of the general public.

Though the futuristic societies portrayed by these movies are incredibly advanced technologically speaking, good and evil still exist and remain the most powerful forces in the universe. In Star Wars the dark or Satanic side of the Force controlled the Emperor and Darth Vader, while the good side flowed strong through Luke Skywalker, Yoda and Obi-Wan Kenobi. At least in the mind of George Lucas, the mystical component of the universe will always prevail over technology in the final analysis.

Science fiction is a popular genre of novels and movies that may potentially generate fabulous returns for the imaginative writer and creator.

On the other hand, fiction science is an abomination, a fairly recent cultural phenomenon where pure science is bastardized in search of the almighty research funding dollar. Fiction science is annoying, intrusive and downright dangerous to that Bank of Knowledge. Fiction science threatens our very liberty and prosperity. It is a politically correct consensus developed by elitists with an agenda to gain power through fear.

The sum total of all humankind has come to know is a Bank of Knowledge. The Bank of Knowledge shapes the decisions we make on both a micro and a macro levels and by extension the course of human civilization. Modern scientific discoveries are only made because modern scientists stand on the shoulders of giants like Newton, Galileo, and Einstein. If future science is based on fallacies added illegitimately to the Bank of Knowledge by fiction science, irreparable harm may result for future civilizations.

ROGUE WAVES

For years, all the computer models said a rogue wave as high as 100 feet was a freak of nature that could only happen once every 10,000 years. As recently as the early 1990s the computer models agreed, so it had to be true.

Pseudo-scientists believed the models in spite of mounting anecdotal evidence suggesting they were wrong. [73]

Statistical evidence such as the fact that over a twenty-five year span more than sixty super tankers sank as a result of unknown reasons was ignored. As Naval architects studied ship disasters between the years of 1969 through 1994, they found more than a third of the super carriers sunk due to sudden flooding actually caused by rogue wave damage.

Until the 1990s, it was generally believed the largest waves that regularly occurred in ocean waters were no more than 60 feet tall. Then the European Space Agency sent a couple of high tech satellites named ERS I and ERS II into orbit. They provided feedback on ocean activity that painted a different picture. The satellites sent detailed information about wave development and tracking from space. This more accurate, freshly obtained data showed well over 400 rogue waves more than 70 feet tall occurred in one small patch of ocean within a period of several months.

The infamous sinking of the Edmund Fitzgerald is now believed to have been caused by two rogue waves that struck the ship in rapid succession. The waves were sighted by a nearby ship moving in the direction of the Fitzgerald shortly before she sank. Navy divers photographed evidence showing the ship had literally snapped in half. Most disturbingly, the sinking was not in open ocean waters, but on Lake Superior in the Great Lakes between Canada and the United States.

In 2001, the cruise ships *Caledonian Star* and *Bremen* were struck by rogue waves off the coast of South America within 10 days of each other. Neither ship sank, but in both cases the waves destroyed the ship's bridge more than 80 feet above the water line. Satellite data showed rogue waves between 90 to 105 feet tall were travelling in the area near where the ships were hit in the same timeframe.

Because of irrefutable evidence, scientists have revised their understanding of rogue waves and new theories are in development. But it took undisputed proof and eyewitness testimony to prompt the change to prevailing scientific opinion and to accept the fact rogue waves can easily reach heights greater than 100 feet.

How difficult will it be to shift the "consensus" opinion of evolutionists to the more plausible hybrid theory?

AN INCONVENIENT TRUTH

Al Gore won an Academy Award and the Nobel Prize for his well-known movie about global warming. It is shown in public schools as a routine part of science classes.

There is only one problem with using the film as a documentary. It isn't one; it's an opinion piece. Gore himself admits a certain amount of exaggeration in the film, which he deemed necessary to raise public awareness of the dangers of global warming. How much exaggeration?

Only several orders of magnitude. The movie depicted floods sweeping across Manhattan as New York City was swamped by a rise in sea levels of nineteen feet. Actual scientific estimates call for a maximum rise in sea level of no more than nineteen inches. The difference in estimates between the claims in Gore's movie and reality is about seventeen and a half feet, making Gore's sea level estimate off approximately twelve hundred percent.

That is a *certain amount* of exaggeration.

Legislation to address global warming is being considered, and emphasis is placed on the urgent need to act without having a clear plan of what to do or what might be accomplished by taking action. Studies by the IPCC itself had to admit the impact of drastic action to reduce carbon emissions like the Kyoto protocol would be negligible at best while the fiscal harm to the global economy would be significant. [74] Sadly, that doesn't stop politicians from attempting to pass these bills because legislation translates to an opportunity for additional taxation.

Recently, leading climate research scientist, Professor Phil Jones of East Anglia University was forced to take an extended leave of absence. Email exchanges between Dr. Jones, Professor Michael Mann of Penn State University, and other leading climate scientists were obtained by hackers and released publicly. The ensuing "Climategate" scandal revealed these scientists conspired to blackball others who were critical of their research data and skeptical to claims of global warming. After the scandal became public, Dr. Jones retracted claims he made about global warming, even admitting global temperatures have cooled slightly since the mid-1990s.

WHY NOT GIVE SCIENCE THE BENEFIT OF THE DOUBT?

Why on earth should we believe the evolutionist more than the creationist if his evidence is not as compelling?

An insurance "study" was conducted that purported to correlate a relationship between deaths of working-age Americans and their lack of health insurance in a transparent attempt to support government attempts to nationalize healthcare. It claimed those "without insurance have a forty percent higher risk of death than their privately insured counterparts." [75]

A number of issues arise with methods used by researchers involved in the study. Study co-author Dr. David Himmelstein is an avid proponent for a single payer health care system and his personal bias apparently influenced the study results significantly.

According to www.junkscience.com chief investigator Steve Milloy, "They interviewed 9,000 people between 1988 and 1994 and asked 'Do you have health insurance?' and if you die at some point in the future, they assume your death was caused by the fact you didn't have health insurance at the time you were interviewed. That kind of stuff is classic junk science."

The article also quoted John C. Goodman of the National Center for Policy Analysis (NCPA), who agreed with Milloy's assessment: "The subjects were interviewed only once and the study tries to link their insurance status at that time to mortality a decade later. Yet the authors have no idea whether the subjects were insured or uninsured, what kind of medical care they received, or even cause of death." [76]

Over the course of history, consensus scientific opinion said the earth was flat, a process called phlogiston explained corrosion and combustion resulting from oxidation, the earth was the geocentric center of the solar system, and every material on earth was composed from one of the four main elementals of earth, wind, fire or water.

Modern scientific theories have not been immune to the need for corrections, either. We were taught there were nine planets in our solar system in science class until very recently.

New information indicates there are either eight or ten planets, depending on your perspective. There is an object, which was discovered to orbit around the sun that is roughly half the size of Pluto. "Eris definitely hurts the case for Pluto being a planet," says Professor of Planetary Astronomy Mike Brown of Caltech. "There is no good scientific way to keep Pluto a planet without doing serious disservice to the remainder of the solar system." [77]

As a result, Pluto was demoted by vote of the International Astronomical Union (IAU) members at a conference in Prague. The solar system now only has eight planets. Sorry, Pluto, but you didn't pass muster.

On the other hand, the periodic table had 109 elements until six new ones were discovered in the last quarter century, resulting in the reprinting of posters for science classrooms all over the world.

There were 30 orders of insects, until a new order called Mantophasmatodea was discovered on the Brandberg Mountain in western Namibia. New studies show man more likely descended from ground dwelling apes than tree dwelling apes. Clothing worn by Ice Age dwellers was not skins and hides but included woven textiles made from plant fibers.

The point is, Science finds evidence and refines theory with additional facts. If the facts support the theory, it is bolstered and conversely the theory becomes weakened if the facts do not. Pseudo-scientists make exceptions for their pet theories, defending them long after logic and hard evidence have refuted them. That fact that debate has been outlawed by legislation is truly mind boggling, worthy of confusing a Bug Blatter Beast even without your towel handy. Why should we trust pseudo-scientific evolutionist theory when a better theory exists?

"To err is human, to forgive divine." Alexander Pope wrote that in an essay on criticism.

We are human and therefore prone to mistake. I can forgive my scientist friends the honest error, provided the debate is reopened in the classroom and evolutionist theory again taught for what it is – a theory. Not a fact.

As long as pseudo-scientists defend the indefensible and argue evolutionist theory has no serious unsettled issues, forgiveness is not merited. Forgiveness requires repentance coupled with a desire to make amends.

12 EXODUS

O Lord my God, When I in awesome wonder,
Consider all the worlds Thy Hands have made;
I see the stars, I hear the rolling thunder,
Thy power throughout the universe displayed.

Refrain: Then sings my soul, My Savior God, to Thee,
How great Thou art, How great Thou art.
Then sings my soul, My Savior God, to Thee,
How great Thou art, How great Thou art!

When through the woods, and forest glades I wander,
And hear the birds sing sweetly in the trees.
When I look down, from lofty mountain grandeur
And see the brook, and feel the gentle breeze. (Refrain)

Quid est veritas?

Those three words might pose the most relevant question one human being ever asked another. Translated from the Latin it means: what is truth?

It was asked by Pontius Pilate of Jesus the Christ in the gospel of John, chapter 18 verse 38. What is truth? Seems like a simple question…but is it?

The truth is comprised of cold, hard facts, free of embellishment. Truth is not like beauty, judged in the eye of a beholder. But as Keats said, the truth is beautiful, "it is all ye know on earth, and all ye need to know." When all lies are stripped away and the truth is all that is left, it is a good thing. The truth really can set you free…but it can also put you in jail for a long time. It all depends on what you've done.

The truth is evolutionist theory is no better and has no more proof than creationist theory. The truth is both theories have flaws. The truth is, though, Divine Evolution makes more sense than either creationism or evolution, it can't be proved either. It just makes more sense.

VERACITY

All men are liars (women and children too, no discrimination here). There are lies of omission and commission, whoppers and little white lies, but they all count. Lies are lies, and we all lie. We lie to protect our loved ones from suffering consequences for their actions, and we lie to incriminate people that we don't like or have made us mad.

We lie to save our own skins from consequences for our wicked deeds, but we're just buying time, we aren't getting away with anything. We lie so effectively that we start to believe our own lies, until we no longer remember the real truth. I'm not saying that I'm not just as guilty, just the opposite is true. In my youth, I lied for self-preservation all too frequently, and I still feel guilty about many of those unpunished falsehoods to this day. I might be the best liar of them all. Don't judge these words by my actions; judge my words by their merit.

When I tell you not to take my word for anything and to check my sources I mean it. Make sure I get quotes and their context accurate and have presented them fairly. I'm not perfect, and I would be shocked to discover I produced flawless work. If that miracle occurred, it would be yet another proof for God because the credit belongs to Him and not me. Without the brain He blessed me with, I could not have written this book.

I do believe that every piece of information I've included is true and accurate, but I've compiled information from a plethora of sources, and they might not all be correct. Anything I found off the Internet was sourced multiple times to prove it wasn't a hoax posting, but I've learned (from my editor) to avoid speaking in absolutes, so I will refrain from absolutely claiming to have verified every source as indisputable.

Show me my mistakes, and I'll be happy to correct them.

QUESTION EVERYTHING

I went about this endeavor with a freely admitted bias toward believing in a God based on my personal experiences, not searching to fill the void of gaps in evolutionary theory. I believe the evolutionist doesn't even begin to effectively explain how things came to be. At the same time I had an honest desire to know more than I did when I started and to see if any evidence had recently been uncovered that should make me rethink my positions on these subjects. After all, my knowledge of evolutionary theory was roughly 30 years old, and given the advances in technology, science, and medicine, I was pretty sure there was a lot I did not know or had forgotten.

There were also new developments about evolutionary theory I needed to learn as part of this process. I wanted to know as much as I could about both sides of the debate in order to be able to respond intelligently to Richard Dawkins. His proverbial "shot across the bow" of calling me delusional motivated me to focus on a comprehensive rebuttal. I suspected I would be rebutting the flawed logical arguments of a prominent author and respected scholar. All I could think was please, allow me to retort.

When people give court testimony, they are asked to give an oath. Please consider this book to be written from the perspective of my being under oath and the information included is true, accurate and complete to the best of my ability, so help me God.

I'm not going to lie and tell you that I have answers to every question or suggest I'm smarter than you. I don't claim that I even know all the questions. I give you my solemn oath that I have honestly conveyed what I believe to be relevant and pertinent information about God, creation and evolution in my possession, obtained from research and personal experience. I do not deny my bias toward belief in God; I gladly confess them all that they may be better understood and known for their honest acquisition.

I will tell you as fairly and impartially as possible, I have researched and documented every relevant fact about evolutionary theory that I could find in my research in order to properly respond to a challenge I felt I had been personally issued by someone I've never met, through a television screen. My current set of beliefs about God and evolution was challenged in such an arrogant and audacious manner by a man named Richard Dawkins, I felt compelled to write this book in response.

Thirty plus years ago I graduated from college after being exposed to and educated on general scientific theory and the basic premise of the theory of evolution. After graduating, I never gave evolution a second thought for years, as it had no relevance on my personal life or business.

I had already concluded that the theory was nonsense. It said humans descended from apes, and if so, why were both apes and humans around in modern times? I was pleased the evolutionists found and solved that dilemma with forked speciation theory, but I found plenty of new concerns to be addressed. I watched Richard Dawkins attempt to address them all under the umbrella of Darwinian natural selection, but that won't fly as evolutionist theory.

If other people wanted to believe evolution that was okay with me, but I ruled it out before I really came to believe in God. When I didn't have a solid faith in God, I couldn't really defend Him as a viable alternative to evolution. I just knew that the people trying to teach me that evolution was indisputable fact were dishonest about the evidence. Had they been able to more intelligently argue the case for evolution, I might have given it a little more credibility.

Evolution lost me for good when my professor and I engaged in an informal discussion of the Big Bang, until I asked, "Where did the matter come from?"

When he responded, "It doesn't matter," I lost respect for him as a teacher and evolution as a comprehensive solution to answer the riddle of how life came to be. I can't even remember which professor or science course it was because my interest in the subject waned as a result of my exasperation.

I wasn't smart enough at the time to articulate a well-researched rebuttal supporting a Creator as a viable alternative. I wish I'd known then the joke about God and pseudo-scientists engaging in a contest to create a human. The joke goes like this: God is challenged by the pseudo-scientists and accepts. He reaches down, grabs a handful of earth, shaping it into human form then breathing life into it. The man gets up and walks away. The team of pseudo-scientists begins preparations in their turn. One grabs a shovel and begins to dig, but God stops him, saying "Wait a minute! Get your own dirt!"

Even the Big Bang Theory requires a densely compressed ball of matter for the explosion to be able to occur. Without God's raw materials, what can we make?

It has never been my intention to become a theologian, and I'm sure writing this book doesn't qualify me as such. I remain skeptical about accepting any claim from anyone at face value. However, there are certain issues that have been settled in my mind for some time like the existence of God and the fact evolution theory is flawed. My research did not dissuade me from those convictions.

I carefully identified and disclosed details of what I learned and the truth of how little we know. I was careful to delineate between conjecture and fact, where reality stopped and the fiction of my imagination (or the evolutionist's) began.

TRUTH FEELS TRUE

Let's face facts…most of the evidence is millions if not billions of years old, which means no living observers exist that can verify its relevance or the accuracy of scientific interpretation of its meaning. Little if any definitive hard evidence exists. The best we can do is examine some very old bones, extract the DNA and compare how similar is the genetic makeup of extinct animals to living ones. One of the few facts we "know" to be true is many examples exist of the previous conjecture and the general consensus have subsequently been proved wrong by more irrefutable evidence. Someone made a guess and someone guessed wrong.

I have no divine revelations from God to impart here to you. Some of what constitutes my beliefs is not founded in the particular religion in which I was instructed. These ideas are more the "evolved" product of my thoughts on God and evolution developed before this effort began and earnestly for the last year or so. Granted, I've spent more time in the last six months thinking about evolution, creation, and the unanswerable questions we all have than I probably did in the previous thirty years, but the existential questions have been revisited on more than one occasion before Colbert interviewed Dawkins.

I have learned a few new things, but nothing that remotely persuades me to believe that God does not exist. According to Dawkins, theoretically evolution is possible because the "miracle" of life only had to happen once. Every vegetative and animal form of life that exists today could be derived from that single anomaly. He doesn't say how or why it is impossible to duplicate by scientific methods, but wants the same faith given the theory that we give to God, but not vice versa.

As I've pointed out, there have been thousands of NDE accounts and even scientific studies on the NDE phenomena, and if only one of them is true, Dawkins' theories about the possibility of a supernatural God have effectively been turned on their head. I have personal experience with the phenomena we call ghosts and believe them to be real and a logical extension of the NDE.

The irony is in the fact that the evolutionist's logic is so convoluted and contradictory when put in context with "the big scheme of things" it is incomprehensible for all intents and purposes. For this precious thing we call life to exist without a divine catalyst, we need an odds-busting collection of coincidental natural events to cause the Big Bang, abiogenesis and speciation to occur that it simply defies all logic without a Creator's involvement.

By some unknown process, we call punctuated equilibrium that same life has to be able to proliferate and differentiate in multiple explosions of growth after being wiped out in multiple mass extinctions. Then the process abruptly stops and cannot be duplicated. The only reason to believe any of this is because you have to, if you don't want to believe in God at any cost.

Life had to survive multiple extinctions with a resilience and apparent foresight combined with an innate intelligence. Life had to be able to generate, replicate, and improve on its own with anticipatory changes to accommodate and adapt to changing environments for future generations. The result is the cumulative effect of appearing to demonstrate intelligent design but it's all an illusion to the eyes of the evolutionist.

Or you can believe you were created and think, My God, how great Thou art, indeed!

BE STILL

I'd like to thank you, the reader, for bringing an open mind to this debate. If you found my arguments compelling, I encourage you to search for God yourself. If you want to find mine, use the Bible. I wish you the best of luck no matter where you look for God. I believe your strongest chance of success is to use the Christian Bible as your guide based on my personal experience.

I refuse to judge another person's faith in God as false. By the same token, I will never accept the religion of another person by force. If Sharia law were ever imposed worldwide, I will be put to death because I cannot deny Christ as my Savior.

I do not judge the Muslim, but I do believe the radical Islamic fundamentalist is the greatest threat to civilization today, much more dangerous than any atheist. A non-religious sadistic murderer like John Wayne Gacy is a serious threat to a community. Religious terrorists may menace society as a whole because they organize and threaten on pretense of serving God. If their God truly mandates such murder and mayhem, he is a false god.

Organized masses of people willing to kill on a whim are obviously more concerning than individual murderers, no matter how gruesome or awful their crimes might be. If the pendulum does change direction and people begin to gravitate back toward belief in God, it will not be due to a miracle cure. Nor will it happen overnight.

Turn off the television, find a quiet spot without distractions, and spend time looking for God. If you seek God with sincerity and perhaps a little desperation, you might be pleasantly surprised by what or Who you find.

People like Richard Dawkins might dwell on ancient atrocities or recent ones committed in the name of God and blame our Creator for them. He mentioned the Spanish Inquisition and the case of a young Italian Jewish boy who was taken from his parents and raised Catholic after receiving last rites then surviving. The story sounded like a sad case of religious people abusing authority even if they considered their intentions noble, but not the fault of God.

I am not worthy to judge God. It is not my place to explain why things happen on earth. But I can tell you what I believe and how I reconcile mortal sadness against supernatural wisdom in my own mind.

I believe God doesn't intervene on a whim to save certain people because He likes some of us better than others. I believe God knows we all will die and what happens after, and if He chooses to intervene or extend our life, it is by His grace according to His prerogative. God knows that we need inspiration now and again, so He performs a miracle in the life of someone who desperately needs it. When that happens, we all can rejoice in their happiness and the hope knowledge of God brings us.

Therein lies one of the mysteries of faith. The change that occurs when one evolves from a skeptical or cynical believer in Christ to a person who really and truly aspires to be more like Him is as obvious as the change to the typical NDE patient post revival.

Richard Dawkins must rationalize every behavior, but he can't understand true Christian behavior because he mistakes its motivation. Dawkins is apparently familiar with insincere Christian behavior, which has adversely affected his view of God. He explains that Christians act because they either fear God and this "eye in the sky" watching their every move, or reciprocal altruism, which translates as "you scratch my back, and I'll scratch yours."

Dawkins illustrates the theory of reciprocal altruism as a lovely example of vampire bats that share regurgitated blood, but when some bats are seen not to reciprocate, they are shunned by their fellow bloodsuckers. What a charming example to illustrate how Dawkins misses the concept of how some people enjoy helping other people with no ulterior motive except they that feel better about themselves from doing a good deed. On the other hand, I can't understand how people can be cruel to other human beings or innocent animals that never did anything to hurt them.

True Christian behavior occurs as the man of one color who sees a woman, a stranger of another color stranded on the side of the road with a flat tire stops and changes the tire for her. Without even giving her his name, he leaves with "God bless" and goodbye, living the story of the Good Samaritan.

It feels good to help other people. It feels good to (platonically) show kindness and compassion to a stranger. It feels bad to hate blindly and stupidly without reason. True Christians don't help people for sex or money. They may enjoy sex and need to make money, but only with their spouses and to pay bills respectively.

You help people who really need help and you feel rewarded though you don't walk away with material wealth gained from the exchange. Spiritual wealth is much more rewarding.

Let there be no mistake, when I pray this book will sell millions of copies I will profit from that exchange, but that was not the reason for writing it. My wife and I both love the Humane Society and the more I can make, the more we can donate to the specific charities we already support.

Evolution is not true. It is not fact. Pure creationism is contradicted by facts as well. The hybrid theory humbly proposed in this book resolves the issues with both traditional theories and provides a seamless and logical alternate theory which accepts the rational facts of evolutionism and creationism and discards the irrational beliefs of both theories. Though the traditional creationist was more right than wrong, it didn't make his arguments all right.

I believe we have a purpose in life. At one time, I thought my purpose was helping animals. Now I believe volunteering with the Humane Society is only part of my purpose, and I'm meant to write. But being an author is not my life's work. I believe writing this one particular book may be. I cannot accept that the literal six days of creation are correct because logic tells me to reject the notion. On the other hand, the thought that a giant ball of matter large enough to create the entire universe coalesced, heated itself up by compressing, then exploding at the speed of light (or faster I suspect) with unbelievably refined precision, and the result was to form the entire universe is also a theory I reject completely when absent the Creator to provide the matter.

Abiogenesis and speciation are less believable than any fable Aesop wrote. Natural selection is a fine theory, until the other three components of evolutionary theory disintegrate without God and no longer remain viable components of the evolutionist argument. The traditional creationist theory does not work appreciably better than evolutionist theory when confined to a strictly interpreted biblical account. We can never know for sure what a day meant to God during creation.

The only reason the evolutionist can justify clinging to his pure theory is a blind and stubborn refusal to accept any possibility of God. If God is impossible, no matter how ludicrous the alternate theory, it must be the answer no matter how improbable.

The Improbability Drive on the spaceship Heart of Gold in The Hitchhiker's Guide to the Galaxy might be able to turn a couple of nuclear missiles into a giant whale and a flowerpot, but it is a fictional invention from the mind of Douglas Adams and isn't real.

We can see creation is real. Stand still for ten minutes, open your eyes and shut your mouth. Watch life pass you by. Birds fly by. People may walk by. Signs of life will be ubiquitous, diverse and abundant. All you have to do is stop and smell the roses, as they say.

I considered beginning this chapter by quoting the lyrics from the hymn Amazing Grace, but How Great Thou Art seemed most appropriate. It calls our attention to the wonders of nature and the miracle of life in all we see. Even a fair-minded skeptic like me can see the beauty and glory of God.

"Be still and know that I am God," in Psalm 46 is both a reassurance to the faithful and a challenge to the cynical. Are you confident enough to truly open your mind, to be vulnerable to the possibility of sensing something greater than yourself?

As we leave this debate of evolutionist versus creationist theory, we take with us a new possibility called Divine Evolution, which addresses the flaws in creationism and evolution not as a compromise solution, but a common sense explanation for life.

There are still the consequences of believing in evolution at the expense of God in society, but that is a topic for another book. We exit the debate for the Promised Land that our faith in God offers. I love the hymn Amazing Grace, and the last two verses seem a most fitting close:

When we've been here ten thousand years
Bright shining as the sun.
We've no less days to sing God's praise
Than when we've first begun.

Amazing Grace, how sweet the sound,
That saved a wretch like me.
I once was lost but now am found,
Was blind, but now I see.

###

ABOUT THE AUTHOR

The author has written a number of articles for the online publication American Thinker and was interviewed on the Dennis Miller radio show. *Divine Evolution* is his first book. He has also written short stories for an anthology about animals and is editing his first detective novel. John is the Atlanta Creationism Examiner for the online new source examiner.com.

John holds a BBA from the University of Georgia and worked as a computer programmer for more than twenty years before becoming a writer. His writing has also been influenced by shorter stints working as a bartender, real estate investor and landlord.

He has been married to wife Lisa for twenty-two years. John is the proud father of two and grandfather of three, as well as pack leader for several wonderful dogs and one crazy cat.

Born in Savannah, John has spent most of his adult life in the northern suburbs of Atlanta. The local color in his writing is equally authentic whether the setting is a Georgia beach, downtown Atlanta, or the Appalachian foothills in north Georgia.

Connect with John online at http://www.southernprose.com/

COVER ARTISTS

The cover photograph is by Kathryn Williams, a retired public school teacher, local politician and restaurant entrepreneur. Kathryn enjoys photographing the incredible beauty of the beaches and marshes of coastal Georgia. She is dedicated to the preservation of our environment and loves living on Tybee Island.

The front cover was designed by Lisa H. Leonard, wife of the author. She is the mother of two and grandmother of three. Lisa is an information technology professional with a specialty in database administration. She enjoys graphic design, gardening, cooking and is active in the local Humane Society.

END NOTES

[1] Jackson, J. "Monkey Scheme." *Keep the Change*. J Jackson. Parodudes, Pittsburgh, 1998. Web. 30 Dec. 2011.

[2] Ripley's Believe It or Not!, *Ripley's Believe It or Not!* Orlando: Ripley Entertainment, 2004. 118. Print.

[3] "Total Perspective Vortex." *The Hitchhiker's Guide to the Galaxy*. 13 Aug. 2008. Web. 30 Dec. 2011.

[4] Adams, Douglas. *Life, the Universe, and Everything*. New York: Harmony Books, 1982. 1. Print.

[5] "Judge nixes evolution textbook stickers ." *Science on MSNBC.com*. MSNBC, 13 Jan. 2005. Web. 30 Dec. 2011. Story.

[6] Berra, Tim M. *Evolution and the Myth of Creationism: A Basic Guide to the Facts in the Evolution Debate*. Palo Alto: Stanford University Press, 1990. 117. Print.

[7] Ibid, page 118.

[8] "Richard Dawkins interview." *The Colbert Report*. The Comedy Channel. 17 Oct. 2006. Web. 30 Dec. 2011.

[9] Genesis Chapters 6-9. *The NIV Study Bible*. Grand Rapids: Zonderman Publishing House, 1995. 14-20. Print.

[10] Berra, Tim M. *Evolution and the Myth of Creationism: A Basic Guide to the Facts in the Evolution Debate*. Palo Alto: Stanford University Press, 1990. 130. Print.

[11] "What Darwin Didn't Know." *National Geographic* 12 Feb. 2009. Print.

[12] Genesis Chapter 1, *The NIV Study Bible*. Grand Rapids: Zonderman Publishing House, 1995. 6. Print.

[13] Job 40:17, Ibid, page 769.

[14] Huda, "Creation of the Universe: Six Days or Long Periods of Time." *About.com Islam*. N.p., n.d. Web. 30 Dec. 2011.

[15] Adams, Douglas. *The Hitchhiker's Guide to the Galaxy*. New York: Harmony Books, 1979. 175. Print.

[16] Crichton, Michael. *Next*. New York: HarperCollins, 2006. 258. Print.

[17] Crichton, Michael. *The Lost World*. New York: Ballantine Publishing Group, 1996. 206. Print.

[18] Ibid, page 207.

[19] Ibid, page 207.

[20] Ibid, page 208.

[21] Ibid, page 392.

[22] Ibid, page 393.

[23] Ibid, page 393.

[24] Berra, Tim M. *Evolution and the Myth of Creationism: A Basic Guide to the Facts in the Evolution Debate*. Palo Alto: Stanford University Press, 1990. 84. Print.

[25] National Academy of Science and Institute of Medicine. *Science, Evolution and Creationism*. Washington DC: National Academies Press, 2008. 24. Print.

[26] Ibid, page 29.

[27] Dawkins, Richard. *The God Delusion*. New York: Houghton Mifflin Books, 2006. 31. Print.

[28] Eastwood, Clint, perf. *Magnum Force*. Writ. John Milius and Michael Cimino. 1973. Warner Brothers. Web. 30 Dec. 2011.

[29] Adams, Douglas. *The Hitchhiker's Guide to the Galaxy*. New York: Harmony Books, 1979. 27. Print.

[39] Siegel, Lee. "The Five Worst Extinctions in Earth's History." *Paleontology*. N.p., 7 Sept. 2000. Web. 30 Dec. 2011.

[31] John 14:6,, *The NIV Study Bible*. Grand Rapids: Zonderman Publishing House, 1995. 1621. Print.

[32] Matthew 7:1, Ibid, page 1448.

[33] Dawkins, Richard. *The God Delusion*. New York: Houghton Mifflin Books, 2006. 52. Print.

[34] Russell, Bertrand. *Why I Am Not A Christian and other essays on Religion and Related Subjects*. New York: George Allen and Unwin Ltd., 1957. 5. Print.

[35] Ibid, page .8

[36] Ibid, page 9.

[37] Adams, Douglas. *The Hitchhiker's Guide to the Galaxy*. New York: Harmony Books, 1979. 59. Print.

[38] Russell, Bertrand. *Why I Am Not A Christian and other essays on Religion and Related Subjects*. New York: George Allen and Unwin Ltd., 1957. 10. Print.

[39] Jung, Karl. *Psychology and Religion: Based on the Terry Lectures delivered at Yale University*. New Haven: Yale University Press, 1938. 33. Print.

[40] Ibid, page 104.

[41] Ibid, page 46.

[42] Ibid, page 48.

[43] Ibid, page 98.

[44] Ibid, page 105.

[45] Ibid, page 73.

[46] Ibid, page 100.

[47] Ibid, page 113.

[48] Ibid, page 114.

[49] Ibid, page 114.

[50] "Fall of the Serial Killer." *Psychic Witness*. Discovery Channel. 12 Jan. 2008. Web. 30 Dec. 2011.

[51] "The Wrong Man." *Psychic Investigators*. Biography Channel. 7 July 2007. Web. 30 Dec. 2011.

[52] "How Dark the Woods." *Psychic Investigators*. Biography Channel. 2 Oct. 2008. Web. 30 Dec. 2011.

[53] "The Other Nostradamus." *Decoding the Past*. History Channel. 18 May 2008. Web. 30 Dec. 2011.

[54] Morse, Melvin, and Paul Perry. *Transformed by the Light: The Powerful Effect of Near-Death Experiences on People's Lives*. New York: Ivy Books, 1992. 187. Print.

[55] Jagger, Mick, and Keith Richards. "Shine a Light." *Exile on Main Street*. Mick Jagger and Keith Richards. ABKCO Records, New York, 1972. Lyrics reprinted by permission.

[56] Morse, Melvin, and Paul Perry. *Transformed by the Light: The Powerful Effect of Near-Death Experiences on People's Lives*. New York: Ivy Books, 1992. 23. Print.

[57] Long, Jeffrey, and Paul Bernstein. "Dutch NDE Study Attracts Worldwide Attention." *International Association for Near Death Studies*. N.p., 29 Jan. 2011. Web. 30 Dec. 2011.

[58] "How many NDEs occur in the United States every day?" *Near Death Experience Research Foundation*. N.P., 13 Aug. 2005. Web. 30 Dec. 2011.

[59] Morse, Melvin, and Paul Perry. *Transformed by the Light: The Powerful Effect of Near-Death Experiences on People's Lives*. New York: Ivy Books, 1992. 194. Print.

[60] Ibid, page 50.

[61] "Richard Dawkins interview." *The Colbert Report*. The Comedy Channel. 17 Oct. 2006. Web. 30 Dec. 2011.

[62] Dawkins, Richard. *The God Delusion*. New York: Houghton Mifflin Books, 2006. 31. Print.

[63] Singh, Anita. "Charles Darwin film 'too controversial for religious America'." UK Telegraph, 11 Sept. 2009. Web. 30 Dec. 2011.

[64] Campolo, Tony. "Katrina: Not God's Will or His Will." *beliefnet: Inspiration. Spirituality. Faith.* . Ed. Tony Campolo. N.p., 6 May 2007. Web. 30 Dec. 2011.

[65] Genesis 28:15, *The NIV Study Bible*. Grand Rapids: Zonderman Publishing House, 1995. 48. Print.

[66] 2 Samuel 18:14, Ibid, page 446.

[67] 2 Chronicles 32:31, Ibid, page 654.

[68] Jagger, Mick, and Keith Richards. "You Can't Always Get What You Want." *Let It Bleed*. Mick Jagger and Keith Richards. ABKCO Records, New York, 1969. Web. 30 Dec. 2011.

[69] Judges Chapter 11, *The NIV Study Bible*. Grand Rapids: Zonderman Publishing House, 1995. 344. Print.

[70] Dawkins, Richard. *The God Delusion*. New York: Houghton Mifflin Books, 2006. 31. Print.

[71] Ibid, page 31.

[72] "Judge nixes evolution textbook stickers ." *Science on MSNBC.com*. MSNBC, 13 Jan. 2005. Web. 30 Dec. 2011.

[73] "Ship sinking monster waves revealed by ESA satellites." esa, 21 July 2004. Web. 30 Dec. 2011.

[74] Michaels, Patrick. "Kyoto Protocol: a Useless Appendage to an Irrelevant Treaty." *Cato Institute: Individual Liberty, Free Markets, and Peace*. Committee on Small Business, U.S House of Rep., 29 July 1998. Web. 30 Dec. 2011.

[75] Carpenter, Amanda. "Junk Science expert sounds alarm on insurance study." *The Washington Times* 17 Sept. 2009. Web. 30 Dec. 2011.

[76] Ibid.

[77] Brown, Mike. "The discovery of 2003 UB313 Eris, the 10th planet largest known dwarf planet." CalTech educational website. Ed. Mike Brown. N.p., 10 Dec. 2005. Web. 30 Dec. 2011.

Made in the USA
Monee, IL
04 November 2026

67375684R00085